Variation Analysis of Ventricular Muscle Condition using Adaptive Algorithmic Paradigms

Zimra

Variation Analysis of Ventricular Muscle Condition using Adaptive Algorithmic Paradigms

First Edition June 2025

Written by Zimra

TABLE OF CONTENTS

CHAPTER NO.	TITLE	PAGE NO.

CHAPTER NO.	TITLE	PAGE NO.

CHAPTER NO.	TITLE	PAGE NO.

CHAPTER 1

INTRODUCTION

1.1 BACKGROUND OF THE PROBLEM AND ITS MOTIVATION

According to the World Health Organization (WHO), cardiovascular diseases have a higher incidence rate across the globe. It is the number one origin of death worldwide and contributes to approximately 30% of all worldwide deaths. Discovering HD prediction is a challenging task that can provide a computerized estimate of the HD level, enabling the simplification of further action. Improving patient outcomes and providing prompt medical interventions depend heavily on early identification and precise prognosis of HD. Consequently, HD prediction has gained significant interest in worldwide healthcare settings. Deep Learning based systems have shown promise in the detection and prognosis of numerous illnesses.

One of the most complex and deadly diseases that affect humans, according to high morbidity and mortality rates, is heart disease, which has a major negative impact on people's quality of life and results in significant financial losses for treatment and monitoring. Artificial Intelligence (AI) holds promise for early detection, diagnosis, and prognosis of health problems. AI has the potential to minimize the severe consequences of heart disease by helping patients get the right medical guidance, help, and treatment from doctors. AI learning approaches can manage massive volumes of Internet of Medical Things (IoMT) data effectively and offer the e-healthcare system highly accurate and efficient real-time cardiac disease detection and prediction.

Cardiovascular disorders are difficult to diagnose due to the numerous contributing risk factors, including diabetes, high blood pressure, excessive cholesterol, irregular pulse rate, and many more. Researchers have been using a variety of data mining and neural network techniques to determine the severity of cardiovascular illnesses in humans. Severaltechniques, including the K-Nearest Neighbor Algorithm (KNN), Decision Trees (DT), Genetic Algorithm (GA), and Naive Bayes (NB), are used to classify the severity of the condition.

The heart is an essential organ in the human body that acts as the engine of the circulatory system. Heart disease has a substantial negative influence on people's quality of life and lifespan, with Coronary Heart Disease (CHD) being one of the leading causes of death worldwide (Rashid *et al.* 2023). Therefore, improving the prevention and treatment of cardiac diseases requires an understanding of the functional mechanisms and characteristics of the heart. In recent years, medical scanners have developed quickly. Examples of cardiac image analysis techniques include single-dimensional heart scans, three-dimensional ultrasound scans, multi-slice cardiac CT, and cardiac Magnetic Resonance Imaging (MRI) (Sundarasekar *et al.* 2021).

1.2 IMAGE PROCESSING

Image processing is the term for any kind of signal processing where the input is an image, such as a picture or a frame from a movie. A picture or a group of characteristics or attributes related to the image can also be the result of image processing. Conventional signal processing techniques are used by most image processing algorithms, which treat the image as a two-dimensional signal.

A. Image

An array, or matrix, is created when square pixels, or picture elements, are arranged in rows and columns to create a digital picture as shown in Figure 1.1.

Figure 1.1 An image is a matrix or array of pixels stacked in rows and columns

An 8-bit grayscale image has an intensity assigned to each picture element as shown in Figure 1.2 which ranges from 0 to 255. Though the word "black and white" is occasionally used to describe a grayscale image, it emphasizes the image's range of grayscale tones.

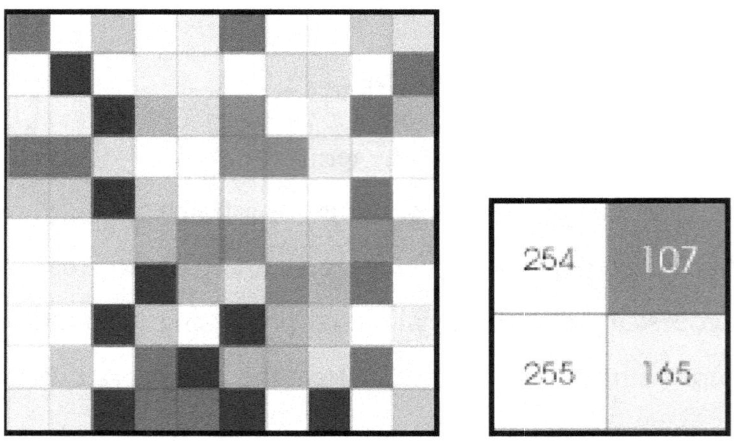

Figure 1.2 The values of each pixel range from 0 (Black) to 255 (White)

The possible range of pixel values is determined by the image's color depth, which in this case is 8 bit = 256 tones or grayscales. The typical grayscale image has 256 grayscales or eight bits of color depth. 256 x 256 x 256 is the result of a 24-bit color depth image, or 8 × 8 x 8 bits of color, roughly speaking, 16 million colors.

Figure 1.3 RGB color images

A composite color image is created by combining three grayscale images with red, green, and blue coloring. An image as shown in Figure 1.3 may have up to 16 million distinct hues. More grayscales can be found in some grayscale photos, for example, 16-bit = 65536 grayscales. It is theoretically possible to create an image with 281,474,976,710,656 gray scales by combining three grayscale photographs. Bitmaps, or pixel-based "images", and vector drawings, sometimes known as line art, are two main categories of digital images. Among the most popular file formats are:

GIF – This is a non-destructively compressed bitmap format with 8 bits (256 colors). It is used primarily for the web and possesses several subpar qualities, the animated GIF being one of them.

JPEG - is a destructively compressed 24-bit bitmap format with 16 million colors that is particularly efficient (i.e., contains a lot of information per byte). extensively utilized, particularly for the Internet and online applications (restricted bandwidth).

TIFF - TIFF is a widely used 24-bit publishing bitmap format. It employs Lempel-Ziv-Welch (LZW) compression, one of the non-destructive compression methods.

PS - PS is an abbreviation for Postscript, a widely used vector file type. It has many substandards and may be difficult to transfer across OS versions and platforms.

PSD - Photoshop's unique format for preserving all of an image's data, including layers, is PSD.

1.3 IMAGE PROCESSING METHODS

The methods available in processing images are:

- Analog Image Processing
- Digital Image processing

1.3.1 Analog Image Processing

"Analog Image Processing" refers to the manipulation of electrical signals that represent images. Television images are the most common example.

1.3.2 Digital Image Processing

The process of processing digital photographs with computer algorithms is known as digital image processing. Consequently, problems like noise accumulation and signal distortion can be avoided and a significantly wider range of algorithms can be utilized to process the input data. Since images are specified over at least two dimensions, if not more, digital image processing can be thought of as a multidimensional system.

Digital processing of the photographs enables the application of far more sophisticated algorithms for image processing, enabling the implementation of techniques that would be impractical for analog approaches as well as more sophisticated performance at simple jobs. While digital processing of the image is the most common variety, optical and analog image processing out are also feasible.

1.4 STAGES OF IMAGE PROCESSING

A. Acquirement - This might be as relaxed as receiving an electronic image The primary tasks include: (a) Mounting (b) Transformation of Color (RGB to Gray or vice-versa).

B. Image Enrichment - This is one of the most straightforward and visually pleasing aspects of spitting image; it also helps reveal elements that are concealed after viewing.

C. Image Refurbishment - This scheme, which is constructed on carefully worked-out or prospects models or image manipulation, addresses aesthetic appeal in images as well as objectivity.

D. Color Image Processing - Color models can be used in digital image processing, dealing with both full and pseudocolor-spitting images.

E. Multi-Resolution Processing and Wavelets - This serves as the foundation for various picture representation levels.

F. Image Determination - It involves the development of several functions to complete this process, primarily dealing with the size or resolution of photographs.

G. Morphological Processing-This area of study focuses on methods for removing image components that help describe and represent shapes.

H. Segmenting the Process - This technique dissects an image into its constituent parts. Autonomous segmentation is the hardest task in image processing.

I. Representation & Depiction - After the output of the segmentation stage, choosing a representation is just one step in the process that transforms raw data into processed data.

J. Object Detection and Recognition - This procedure assigns an object a label by utilizing its descriptor.

1.5 IMAGE PROCESSING TECHNIQUES

The different methods for image processing include:

- Image Representation
- Image Preprocessing
- Image Refurbishment
- Image Segmentation
- Image Enrichment
- Image Determination
- Image Interpolation

1.5.1 Image Representation

Typically, quantized images are shown as collections of pixels that encode color and brightness data in a matrix format. Contour lines form the basis of an alternate model. The complete bitmap image can be easily retrieved using a contour representation. It has mostly been applied to image data compression. The goal is to encode, at levels larger than or equal to, the borders of connected regions of pixels at each level. From those boundaries, an original image can be easily recreated. There are algorithms for computing contour representations that are sensitive to output.

1.5.2 Image Refurbishment

The goal line of image refurbishment is to "undo" or "compensate for" the flaws that cause an image to degrade. Degradation comes in many forms, including as sound, signal distortion, and improperly focused cameras. It is possible to precisely forecast the blurring function in scenarios like signal distortion and "undo" the blur to restore the original image. When sound distorts an image, the best we can hope for is refurbishment of the damage caused by the sound.

1.5.3 Image Segmentation

The process of segmenting a picture is to make its representation easier to comprehend and analyze. Image segmentation is mainly used to find objects and boundaries (lines, curves, etc.) in pictures. To be more precise, image segmentation is the process of assigning a label to every pixel in a picture so that those pixels have similar visual characteristics. The results of image segmentation are either a set of segments that combined cover the entire image or a set of contours that are removed from the image (see edge detection). Every pixel in a region is comparable with respect to a certain property or calculated feature, such as color, texture, or intensity.

Segmentation is an essential step in order to extract information from complex medical pictures. The usage of segmentation is common in the medical industry. Image segmentation's main objective is to partition a picture into distinct and mutually exclusive portions. Every area of interest needs to be physically connected, and each pixel needs to meet a set of standards. Examples of often used homogeneity criteria are values for surface normal, surface curvatures, range, texture, intensity, and color. In the past, a large number of researchers in the domains of soft computing and medical imaging have carried out in-depth studies related to picture segmentation. Image segmentation methods can be divided into groups based on threshold levels, edge detection, region or surface growth, classifiers such vector quantization and Hierarchical Self Organizing Maps (HSOM), and so on feature path clustering.

The process of segmenting a picture involves breaking it up into sections, each of which is made up of neighbouring pixels with comparable attributes like color, intensity, etc. To identify a pattern, two classifications are available: supervised classification and unsupervised classification. Supervised classification is the process of determining a mathematical function that can forecast a class's membership based on incoming data. Unsupervised classification is the process of classifying pixels based on their numerical spectral properties without the need for direct human instruction.

1.5.4 Image Enrichment

The process of improving the quality of a digitally stored image by modifying its contrast, brightness, or other elements. More advanced image enrichment software comes with a variety of filters to alter photos in various ways. Software programs made expressly for manipulating images are called image editors. Color, hue, and brightness imbalances can often be corrected automatically with basic image enrichment techniques found in camera or computer image editing programs. Additional image editing features are also often offered, such as cropping automatically, zoom features, sharpness tweaks, and red eye removal. Because they typically happen automatically or may be started with a single button press, mouse click, or menu selection, these are referred to be automated.

1.5.5 Image Determination

The goal of image compression is to reduce superfluous information in the image data to optimize its storage or transmission. It is possible to compress images losslessly or with loss.

- Lossless compression is the best option for historical contexts and is often used for technical drawings, cartoons, medical images, and clip art. This is because, especially when used at low bit rates, lossy compression algorithms produce compression artifacts.
- In particular, lossy algorithms work particularly well for natural images such as photographs, when a slight (sometimes unnoticeable) loss of quality is acceptable but a considerable bit rate decrease is required. The term "visually lossless compression" refers to lossy compression that produces little perceptible changes.

1.5.6 Image Interpolation

The procedure of resizing an ordinal spitting image is termed spitting image interpolation, sometimes known as image scaling.

The process of scaling is complex and requires balancing efficiency, smoothness, and sharpness. An image appears "soft" as its size increases because the individual pixels are more visible. On the other hand, an image's smoothness and perceived sharpness will generally improve with reduction. In addition to resizing the image to fit a smaller display area, thumbnails are typically created by reducing the image size, also known as subsampling or downsampling. Making smaller graphics suit a larger screen in full screen mode usually involves widening a spitting image (up selection or interpolation). For instance, when you "zoom" in on an image, the quality of the image invariably degrades since you cannot find any more information in the image than what is currently there.

1.6 MORPHOLOGICAL OPERATIONS

Digital images are processed based on their shapes using a variety of methods referred to as morphological operations. In a morphological process, each pixel in the image has a value that matches every other pixel in the vicinity. By choosing a neighborhood pixel's size and form, a morphological operation that responds to specific shapes in the input image can be built. By applying a Matlab-named structuring element to an input image, morphological procedures produce a similarly sized output image.

1.6.1 Morphological Operation Types

Dilation: Pixels are further added to the boundaries of things due to dilation. To apply a dilation operation, we traverse the structuring element across the image object. It's able to fix cracks and breaks in objects.

The following formula is used to determine the output pixel values.

Output pixel = 1 {if HIT}

(output) pixel = 0 {otherwise}

Erosion: Pixels on object borders are removed due to erosion. To execute an erosion operation, we move over the picture object while traversing the structuring element. It may remove extrusions and split apart joint items.

The following formula is used to determine the output pixel values.

Output pixel = 1 {if FIT}

(output) pixel = 0 {otherwise}

Open: The opening procedure dilates the deteriorated image after eroding it using the same structural element for both procedures.

Close: The closure operation dilates an image and then erodes the dilated image using the same structural element for both processes.

How many pixels are added to or deleted from the object in the image depends on the shape and size of the structuring element that was used to process the image. The condition of every particular pixel in the output image during the morphological dilation and erosion processes is determined by applying a rule to the associated pixel and its neighbours in the input image. The rule that handles the pixels defines the morphological operation as dilation or erosion.

1.7 EDGE EXTRACTION OF AN OBJECT

Determining the boundaries of a picture is a crucial step in gathering data and comprehending its characteristics. It is the initial pre processing step that displays the properties of the image. The researcher can obtain data from the image with the aid of this procedure. By taking the actions listed below, we can extract an object's boundaries.

1. Use the erosion process to create a picture (E), which will cause it to slightly shrink. It is possible to adjust the structural element's kernel size appropriately.
2. Take picture E and subtract it from the original image. We determine the boundaries of our object by completing this step.

1.8 MAGNETIC RESONANCE IMAGING (MRI)

The main purpose of Magnetic Resonance Imaging (MRI) in medicine is to see how the body is constructed and functions. A cutting-edge imaging technique called Magnetic Resonance Imaging (MRI) makes it possible to observe the body cross-sectionally with previously unheard-of tissue contrast. The amount of knowledge on the use of magnetic resonance imaging in evaluating pathologic disorders of the ankle, foot, and brain has increased dramatically during the last fifteen years. It has quickly become a recognized technique for medical imaging of disease processes in the musculoskeletal system, particularly the foot, and brain, and it doesn't utilize any ionizing radiation. It offers fine-grained pictures of the body in any direction. MR is particularly helpful for neurological, musculoskeletal, cardiovascular, and oncological imaging since it has far more soft tissue contrast than Computed Tomography (CT). In contrast to CT, it aligns the body's hydrogen atom magnetization using a strong magnetic field instead of using ionizing radiation. To determine the concentrations of various metabolites in bodily tissues, magnetic resonance spectroscopy is utilized. The isotope being "excited" by the MR signal exhibits a spectrum of resonances that correlate to various chemical configurations. This signature is used to offer information on tumor metabolism and to diagnose specific metabolic diseases, particularly those that impact the brain.

More and more people are realizing that MR imaging is the best technique for evaluating brain pathologies. Any tissue plane can be used to obtain digital representations of tissue properties through magnetic resonance imaging. Nuclear Magnetic Resonance (NMR), a spectroscopic method that scientists employ to gather microscopic chemical and physical details about molecules, is the foundation of Magnetic Resonance Imaging (MRI). Due to the negative connotations of the word nuclear in the late 1970s, Magnetic Resonance Imaging (MRI) was chosen for the technique instead of nuclear magnetic resonance imaging (NMRI). MRI was initially developed as a tomographic imaging method, which creates an image of the NMR signal through a tiny section of the body. Tomographic imaging has been improved from MRI to a volume imaging method. (Paragraph Repeated)

More and more people are realizing that MR imaging is the best technique for evaluating brain pathologies. Any tissue plane can be used to obtain digital representations of tissue properties through magnetic resonance imaging. Nuclear Magnetic Resonance (NMR), a spectroscopic method that scientists employ to gather microscopic chemical and physical details about molecules, is the foundation of Magnetic Resonance Imaging (MRI). Due to the negative connotations of the word nuclear in the late 1970s, Magnetic Resonance Imaging (MRI) was chosen for the technique instead of nuclear Magnetic Resonance Imaging (NMRI). MRI was initially developed as a tomographic imaging method, which creates an image of the NMR signal through a tiny section of the body. Tomographic imaging has been improved by MRI to a volume imaging method.

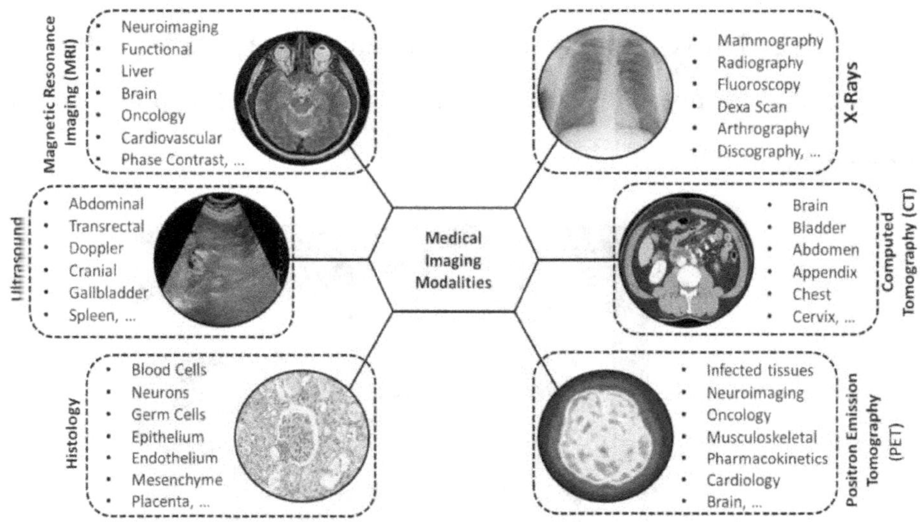

Figure 1.4 Types of medical imaging modalities

The MRI modality is regarded as the gold standard among them. Since cardiac MR imaging doesn't involve ionizing radiation, it is safe (Ripley *et al.* 2016). MRI is utilized to assess the cardiovascular system's structure and function non-invasively. Conditions for which it is carried out include coronary heart disease, dissection, aneurysm, coarctation of the aorta, congenital heart disease, cardiomyopathies, and valvular heart disease. It is also useful for examining pulmonary veins. Using myocardial wall motion and function, the existence and degree of myocardial edema, ischemia, and scarring, as well as structural characterization of the heart, CMR regulates the presence, acuteness, and severity of heart disease. Its flexibility is unique to other numerous imaging methods. It provides not only accurate anatomical data but also efficient data that helps in finding patients at risk.

CMR imaging is widely used for the evaluation of left and right ventricular volumes, ejection fraction, myocardial mass, and wall thickness as well as analyses of wall motion abnormalities (Grothues *et al.* 2004). Precisely following the LV and RV contours is necessary to calculate these parameters. When evaluating, short-axis cine pictures that show the whole left and right ventricles from base to apex are given special consideration. The short-axis image (slice) of the left ventricle is shown in Figure 1.5. Approximately 20 to 25 frames are captured for a whole cardiac cycle. A standard inspection could call for 12 to 15 slices or 300 to 400 photos in total.

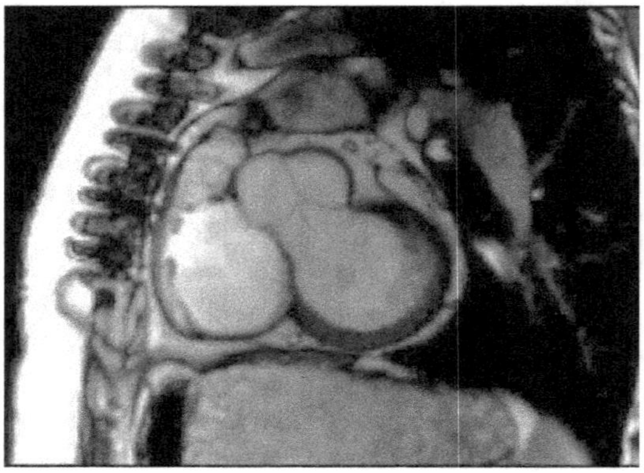

Figure 1.5 Short-axis view of the left ventricle

Cardiovascular Magnetic Resonance Imaging (CMRI) scans are frequently used to obtain clinical characteristics of the Left Ventricle (LV), including ejection fraction and LV volumes. After receiving thecardiac images, the clinicians segment or trace the contours of the ventricular in the ED (End-Diastolic) and ES (End-Systolic) stages. This post-processing step requires 10–20 min for a trained clinician. This manual procedure takes too much time, is tiresome, and is subject to humanerrors (Bai *et al.* 2015). Furthermore, medical practitioners find it challenging and time-consuming to manually segment cardiac illnesses and evaluate such functions. The major objective of ventricular segmentation is to trace their contours and estimate the diagnostic measures. Though the diagnostic measures could be estimated even without segmentation, the contours are much needed for clinical experts. However, automated segmentation is difficultdue to the shrinkage of ventricles Figure 1.6 (Choromanska *et al.* 2015).

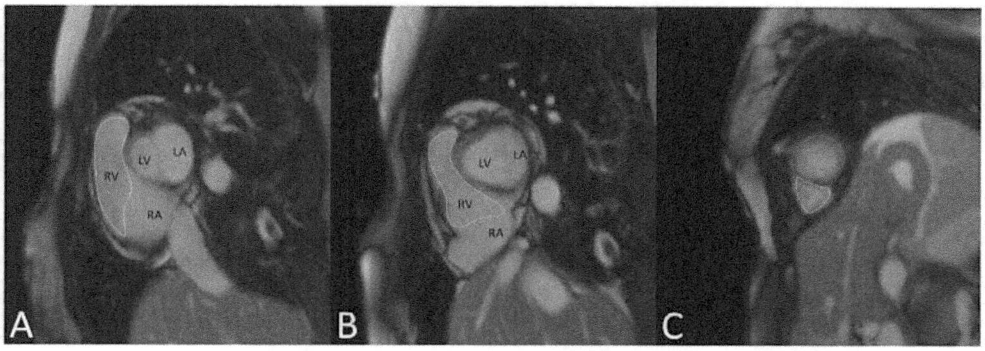

Figure 1.6 Difficulties in LV segmentation (LV: Left Ventricle, LA: Left Atrium, RV: Right Ventricle, RA: Right Atrium), in (A) and (B) it is difficult to delineate the LV from LA, in (C) the LV shrinks towards the apex making the segmentation complex

Therefore, it is essential that segmenting data effectively and economically will help medical professionals make a more accurate diagnosis. Considering both LV and RV from the cardiac images, the LV is responsible for pumping the arterial blood to the artery and subsequently to the systematic connections. The LV has an elliptical structure encircled by the myocardium; whose wall thickness ranges from 6 to 16 mm. The RV, on the other hand, has a complex structure and 3–6 times thinner wall size thanthe LV. Based on the functionality the LV is more imperative than RV, hence the research community focuses on the LV segmentation instead of RV (Cong *et al.* 2019).

The following are the open challenges in automated LV segmentation (Queirós *et al.* 2014; Tavakoli & Amini *et al.* 2013; Petitjean & Dacher *et al.* 2011).

- Pixel intensity range overlap between cardiac components and neighboring background objects.

- The shape of the cardiac objects varies across slices and phases.
- Greater imbalance exists between the amount of background pixels and the pixels belonging to the object.
- The contours are fuzzy and irregular. Variability due to different modalities and scanners.
- Heterogeneities in the brightness of LV cavity due to blood flow.
- Partial volume effects in apical slices due to the limited resolution of cardiac MRI.
- Considerable variability in shape and intensity of the heart chambers across patients.
- And the inherent noise coupled with cine MRI.

Although research studies are addressing many issues, still the segmentation algorithms have to be enhanced to achieve better clinical decisions (Cong *et al.* 2020). Most of the existing methods require a priori knowledge such as ventricular structure, spatial relationships, and their surrounding objects. Developing a segmentation system with prior knowledge of the domain would always improve the accuracy and robustness, but only the strong or appropriate priors should be identified and used, otherwise, the segmentation might suffer with over-fitting and under-fitting issues.

Automatic ventricular segmentation with short-axis cine CMR image has been intensively carried out over the last two decades. The research contributions are motivated by various challenges launched by several research communities, such as the Sunnybrook Cardiac MRI left ventricular segmentation challenge (Tan*et al.* 2018), LV segmentation Dataset and Challenge (LVSV), and Automatic Cardiac Diagnosis Challenge (ACDC) (Bernard *et al.* 2018). Automatic cardiac MRI segmentation has been studied with pixel classification, image-based approaches, deformable methods, and aspect-based models. However, these approaches suffer from low robustness, user intervention, and bias due to the small training dataset (Avendi *et al.* 2016). At the same time, the learning-based approaches such as Random Forests (RF), Conditional Random Field (CRF), and Restricted Boltzmann Machines (RBM) could have been considered, but due to the complexity of parameter estimation, they are still computationally challenging, and their stability is not assured (Dreijer *et al.* 2013).

1.9 DEEP LEARNING

Deep artificial neural networks are models used in deep learning. Neural networks consist of multiple hidden layers, an output layer, and an input layer. Artificial neural networks serve as the foundation for a collection of statistical machine learning techniques known as "deep learning," which are used to learn feature hierarchies. Deep learning is implemented using deep networks, which are neural networks with several hidden layers.

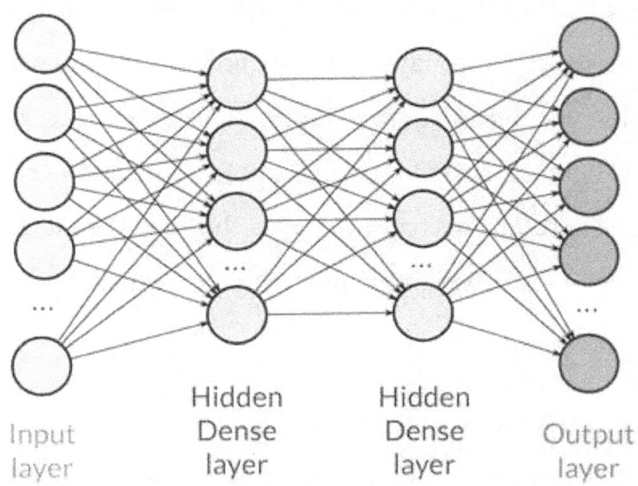

Figure 1.7 Deep learning's architecture model

In Figure 1.7 given above, an input layer is the first layer. Following that, these input layers will identify local contrast patterns, which implies they will distinguish based on hues, brightness, etc. The image feature will then be determined by the first concealed layer. It will then correct those features on the appropriate image template. Therefore, it will identify the proper image in the second concealed layer before sending it towardsthe output layer. Similarly, extra hidden layers can be additional towards answering extra intricate puzzles, like identifying a specific type of image with light or large complexions. We can therefore tackle complex problems as and when the number of hidden layers increases.

1.9.1 Deep Learning Network Types

A. Feed Forwarding Neural Network

Nodes are kept from creating cycles by an artificial neural network, often known as a feed-forwarding neural network.

This kind of neural network has all of its perceptron organized in layers, with the input layer generating output and the output layer receiving input. Because there is no communication between the hidden levels and the outer world, they are known as secret layers. Every node in the subsequent layer is linked to every single perceptron in that layer. It is concluded that each node is fully connected. There are no visible or invisible connections between the nodes on the same layer. There are no back-loops in the feed-forward network. To lower the prediction error, the weight values can be adjusted via the back propagation method.

B. Recurring Neural Network

The recurrent neural network is another type of feed-forwarding network. Here, after a predefined period, each neuron in the hidden layers receives an input. Most of the time, historical data from earlier iterations is retrieved by the Recurring Neural Network. To anticipate the word that will follow next in a particular sentence, for example, one needs to know what words have come before. In addition to sharing the length and cross-wise weights, it processes the inputs. It stops the model from getting bigger as the input gets bigger.

C. Autoencoding

An autoencoding neural network is another kind of unsupervised machine learning algorithm. Simply said, there are fewer hidden cells in this instance than input cells. On the other hand, there are the same number of input and output cells. An autoencoding network is trained to show the output similarly to the fed input to force AEs to find common patterns and generalize the data. The autoencodinghandles most of the smaller representations of the input. It makes it easier to put together the original data from compressed data. This algorithm is really simple; all it needs is for the output to match the input.

- **Encoder:** Reduces the size of the input data.
- **Decoder:** Put the compressed data back together.

D. Convolutional Neural Network

Convolutional neural networks are a special kind of neural network that is mostly used for image clustering, object recognition, and picture categorization. Hierarchical image representations can be constructed unsupervised with the help of DNNs. Deep convolutional neural networks are the most popular neural network type to attain optimal accuracy.

The sole distinction between fully connected feedforward and convolution neural networks is in the connections between their two surrounding layers In contrast to the latter, which has every node connected between the two neighbouring layers, the former only has a portion of its nodes connected. The main problem with using a feedforward neural network that is completely linked is that there are too many parameters for the network to handle. Generally speaking, raising the parameters will result in overfitting issues in addition to slower calculation speeds. Reducing the number of parameters in neural networks successfully requires more intelligent neural network structures. Thus, convolutional neural networks were proposed as a way to accomplish this goal.

The convolutional and pooling layers, which are the two kernel layers of convolutional neural networks, are as follows:

- **Convolutional Layer:** Only a small portion of the previous layer is fed to each node in the convolutional layer; this tiny region is typically 3 × 3 or 5 x 5. The convolutional layer produces a higher abstraction of feature representation by attempting to thoroughly examine each tiny region of the neural network.
- **Pooling Layer:** Usually, the convolutional layer comes after the pooling layer. The number of parameters in the neural network can be decreased by the pooling layer since it can successfully reduce the size of the matrix from the preceding convolutional layer. Consequently, the application of pooling layers can avoid overfitting issues while simultaneously expediting computation. Based on the distinct connection modes of the various convolutional layers, there are two main categories of convolution neural network topologies. One approach involves joining various convolutional layers in a series, like LeNet-5, AlexNet, and ZFNet. The other involves joining various convolutions in parallel, like Inception and its successors.

1.9.2 Applications of Deep Learning

Among the applications of deep learning in MRI image processing and analysis include image recognition, image registering, picture segmentation, and image classification. These may all be expressed as feature representation problems, which can then be successfully resolved by identifying an efficient set of features using deep learning techniques.

A. Image Recognition

A major component of computer-aided detection algorithms is picture recognition. The first step is to identify the pertinent tissues; next, it measures the tissues and determines whether they create lesions. Some deep learning methods that have been proposed for MRI image recognition include the following:

A deep learning layered sparse autoencoder model is used to identify the organs from an unusually complex dataset where it is difficult to discern the labels of the samples. In this work, the stacked sparse autoencoder model is the result of stacking several unsupervised feature learning layers that were trained via greedy methods. After that, a pooling strategy was used to compress the characteristics of progressively larger input regions, producing a part-based model for multiple organ detection in MRI images.

The automated detection of lacunes with a presumed vascular origin is achieved using a two-step deep convolutional network technique. First, a fully convolutional network was used to identify the initial candidates. Next, a 3D convolutional network was employed to reduce false positives. The geographical information is essential in this instance for identifying possible tissue. As a result, by incorporating multiple scale analysis and a combination of explicit location data into the neural network, detection performance was further improved by contextual information.

An autonomous 3D convolutional network technique is introduced for MRI image-based Cerebral Microbleed (CMB) identification. The 3D convolutional network was employed to extract high-level features for CMBs using a data-driven methodology, effectively encoding the spatial contextual information from MRI images. The traditional sliding window strategy employed by the 3D convolutional network resulted in a rather high computing cost for the method's CMB detection. A two-phase cascaded 3D fully convolutional network design was proposed to improve CMB detection performance. The 3D fully convolutional network was utilized to efficiently recover potential candidates, which were then employed to further distinguish challenging mimics from CMBs.

B. Image Registering

Image registering is the process of matching and superimposing two or more photos acquired under different conditions (such as illumination, position, and angle) or at different times, using different sensors (like imaging equipment). The general process for image registration is as follows:

- Two photos features are extracted to obtain the features.
- The similarity measure is used to find the feature pairs.
- The parameters for the image space coordinate transformation are obtained by matching pairs of features.
- The parameters for the image space coordinate transformation are used to perform image registering.

Picture registration has been widely used in medical image processing. Its main objective is to combine several medical photos into a single image that has all of the information it contains, providing users with access to a multitude of sources for medical diagnostic information. Therefore, to perform medical image registration, accurate and effective correspondences between the two images must be constructed. In general, maximizing the feature pair similarity can be used to indicate the correspondences between two images. Recent advances in neuroimaging techniques have led to the emergence of several innovative modalities that enhance the precision of disease diagnosis and therapy.

Picture registration operations, which combine data from many modalities, are therefore required. Many learning-based image-registering approaches have been developed to select the most relevant features that would lead to the corresponding detection between samples with significant changes. However, a significant disadvantage of most learning-based picture registration methods currently in use is that they require a large number of detected correlations to be identified during the instruction process. This issue was addressed by Wu *et al.* (2020) unsupervised deep learning system, which extracts the optimal image features for MRI image registration. In this work, a stacked convolutional independent subspace analysis network was first developed to learn the hierarchical representations of patches using MRI brain images. The stacked convolutional independent subspace analysis network consisted of two layers: (1) The first layer extracted the low-level feature representations of patches from MRI brain images; (2) the second layer extracted the hierarchical representations. The hierarchical representations were then used to complete the correspondence detection phase of the picture registration process. Later, (Wu *et al.* 2020) proposed a deep learning framework that is unsupervised and may be used by the same team to learn hierarchically.

MRI scan representations to record the picture. A stacked autoencoder and a convolutional network were incorporated into the unsupervised deep-learning framework. The 3D photo patches were the inputs used to train a convolutional network-based stacking autoencoder. The bulk of the stacked autoencoder employed in this study consisted of the encoder and decoder networks, from which the former was used to learn the low-dimensional features from 3D picture patches and the latter to recover the 3D image patches from the learned low-dimensional features. However, if the inputs are really large, using the stacked autoencoder directly will result in a very high computational cost. This work reduces the dimensionality of the initial features and the computational cost of the stacked autoencoder by learning the translational invariant using a convolutional feature.

C. Image Segmentation

An essential component of contemporary clinical practice and research in medicine is automatic tissue segmentation in MRI images. Many medical image segmentation challenges have stimulated the development of automatic segmentation algorithms, such as cardiac MR Left Ventricle (LV) segmentation (http://smial.sri.utoronto.ca/LV_Challenge/Home.html). Numerous deep-learning techniques have also been put out to segment different tissues in magnetic resonance imaging. Measuring the cardiac ventricles, comprising the LV and RV, is crucial for the clinical assessment of cardiac architecture and functions, including ventricular volume, wall thickness, and ejection %.

Consequently, the precise and automated division of the heart ventricle is alsoessential. A plethora of deep learning techniques have been proposed recently for cardiac ventricle segmentation. For instance, (Avendi *et al.* 2015) presented an approach to separate the LV from MRI images, combining deformable models and deep learning architectures to accomplish this objective. There were three essential steps to the method: (1) Using stacked autoencoders to infer the LV's shape; (2) Using convolutional networks to predict the LV's location from MRI images; and (3) Using the inferred LV shape to carry out the LV's final segmentation by a deformable model. a hybrid approach for automated LV segmentation from MRI images that combines a level set algorithm and deep learning architecture. Three steps are also included in the method: (1) A deep network of assumptions to use magnetic resonance imaging (MRI) to locate the left ventricle (LV); (2) An additional deep belief network to define the boundaries between the heart and the epicardium; (3) To complete the final LV segmentation, the boundaries between the epicardium and the heart as well as the predicted location of the LV were incorporated into the distance regularized level set approach. In addition to the tissues listed above, deep learning algorithms were used to segment the MRI images of several tissues, including multiple sclerosis, the prostate, the striatum, the tibial cartilage, the abdominal fat tissues, and the anterior visual pathway.

D. Image Classification

Image categorization plays a major role in both automatic disease diagnosis and cognitive recognition, including distinguishing between diseases of varied severity and different cardiac activity. Numerous deep learning techniques have been put forth to carry out image categorization tasks in MRI scans.

1.9.3 Deep Learning Methods Applied to MRI Scans

Based on the generic deep learning methods discussed above, various MRI scan-specific deep learning methods have also been developed recently. Here is a little introduction to them.

- **BrainNet:** Using Tensor Flow as a base, this tool trains deep neural networks to discriminate between GM and WM in brain MRI images.
- **LiviaNet:** The program, was also created using Theano as a basis, and uses MRI pictures to separate subcortical brain regions to train 3D fully convolutional neural networks.
- **DIGITS:** This method was likewise created the train precise deep neural system sused for applications like tissue identification, image segmentation, and classification. DIGITS, for instance, is used to successfully forecast Alzheimer's disease using MRI pictures.
- **ResNet CNNMRIADNI:** This program was created to do AD classification of MRI images to train enduring and also simple convolutional neural systems.
- **Mrbrain:** Convolutional neural networks are trained by this algorithm to predict human age from MRI data.
- **DeepMedic**: It's a program made to train multi-scale 3D convolutional neural networks so they can identify different parts of the brain from MRI pictures. It was created using the Theano framework. Additionally, this technology won the ISLES 2015 competition and demonstrated exceptional performance in brain lesion segmentation tasks.

1.10 LEFT VENTRICLE SEGMENTATION USING CNN

Recently, deep learning-based segmentation automation approaches have reported promising results. Bai *et al.* (2018) and Duan *et al.* (2019) based on this context FCN, CNN + U-net, 2D-UNet, and U-Net models have been established to analyze the global and local functions of the cardiac. Typical deep learning classifiers like AlexNet, VGGNet or GoogLeNet, read an image and output the class that it belongs too. This is achieved by processing the input image through the sequence of functions like convolution, pooling, activation and learning as shown in Figure 1.8.

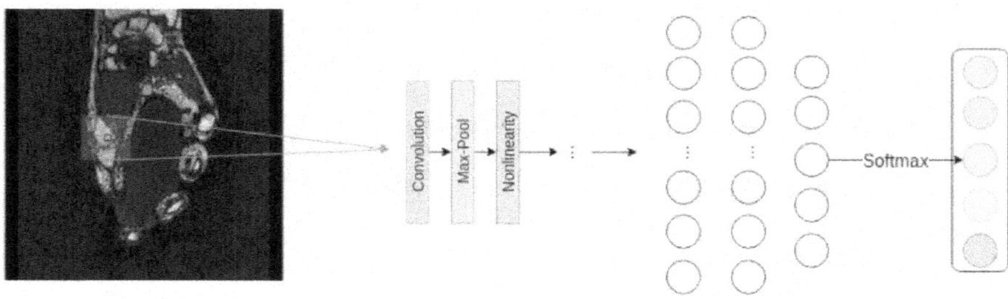

Figure 1.8 Convolutional neural network architecture for image segmentation

Hand-written character recognition is an example for a classification problem, where an image of a hand-written symbol will be the input and the classifier has to recognize as the correct character. Segmentation problem, on the other hand, requires the classifier has to label each pixel of the input image. The basic idea of deep learning based image segmentation is to applying a classifier to patches of the input in sliding window fashion, where the performance highly depends on the patch size. One limitation here is that the patches can learn only the local features; however, the efficient segmentation requires global descriptors too. In CNN architecture, fully connected layers are responsible for producing final classification results. For the segmentation problem, the classifier output should be threshold for the binary segmentation of each pixel. Lin *et al.* (2013) proposed a Fully Convolutional Neural Network (FCN) that is an end-to-end architecture, where the fully connected layer is ignored. Long *et al.* (2015) further improved the FCN by adding a 1×1 convolution. While replacing fully connected layers with convolutions allows for multiple pixels to be predicted efficiently and simultaneously, the resulting outputs typically have a very low resolution, which could be resolved by up sampling. Hence the FCN based medical segmentation is performed by extracting a patch and doing inference by sliding the CNN across all pixels of the network, predicting one pixel in each forward pass and training an FCN on full images or large sections of images. However, deep learning based segmentation approaches are expensive in terms of time and resources, and the choice of architecture and hyper parameters are to be optimally chosen for efficient segmentation results (Mortazi & Bagci *et al.* 2018). Also, it is expected that the segmentation system should be able to handle populations of diverse patients with various ventricular structures. This objective indicates that the automation system should be trained with more number of samples Duan *et al.* (2019).

On the other hand, some of the deep learning-based detection algorithms are also introduced. In the WE-SAJ model, (Wang *et al.* 2022) the wavelet entropy was used for feature extraction, two-layer FNNs were also used for classification, and the adaptive Jaya algorithm was used as a training algorithm. By utilizing the Particle Swarm Optimization Self-Tuning Convolution Neural Network (PSTCNN), (Wang *et al.* 2023) it may be possible to minimize the need for human intervention in the diagnosis. Moreover, the optimization method can select hyperparameters selectively to achieve a near-optimal result by stabilizing the set of parameters.

The fundamental concept of deep learning-based picture segmentation is to categorize input patches in a sliding window manner, with a strong dependence on patch size on performance. One limitation here is that the patches can learn only the local features; however, the efficient segmentation requires global descriptors too. In addition, the deep learning-based segmentation approaches are costly in terms of time and resources, and the choice of architecture and hyperparameters are to be optimally chosen for efficient segmentation results (Mortazi *et al.* 2018). Also, it is expected that the segmentation system should be able to handle populations of diverse patients with various ventricular structures.

This objective indicates that the automation system should be trained with a greater number of samples. (Wu *et al.* 2020; Hu *et al.* 2019; Shaaf *et al.* 2022 and Upendra *et al.* 2019) based on the existing techniques for segmentation of LV from the cardiac images depend on the prior structure and the features are unstructured. These features are applied to the deep learning-based segmentation task which gives over or under-segmentation results and gives poor accuracy.

Cardiac ultrasound images play a crucial role in evaluating physiological indicators of the heart and diagnosing conditions utilizing deep learning (Fenil *et al.* 2019). The ease of operation, reliability, and practicality of cardiac ultrasound technology make it a potent non-invasive method for conducting comprehensive assessments of cardiac function. The systemic blood supply is managed by the left ventricle. Acquiring measurements such as left ventricular end-diastolic volume, left ventricular end-systolic volume, left ventricular ejection fractions, and stroke volume aids in comprehending alterations in the left ventricle (Steinhaus *et al.* 2019 and Halliday *et al.* 2021). This process supports the establishment of a measurable approach to prevent and address heart ailments, thereby mitigating the risks and mortality associated with cardiovascular ailments. The pivotal point of the heart is the left ventricle and signs associated with it play a crucial role in diagnosing heart disease (Hassan *et al.* 2019). Therefore, obtaining precise information is essential for subsequent prognosis. However, the morphology of the left ventricle varies across these views, necessitating the classification of these views and subsequent left ventricle detection. This detection process significantly reduces the time doctors spend searching for pertinent information within extensive echocardiographic data (Pastore *et al.* 2020 and Joyce *et al.* 2022). The manual description is a time-consuming process and is susceptible to both individual observers and between different observers. Depending on the LV segments using short-axis MRI scans, left ventricular medical variables are commonly gathered (Curiale *et al.* 2019). Medical professionals find that manually segmenting data to evaluate these functions takes a lot of time and effort when diagnosing heart diseases. Nevertheless, repeatedly segmenting these images by hand reduces the reliability of the diagnosis (Ribeiro *et al.* 2022).

To overcome unstructured features this research develops a unique Edge and Shape feature-based Fully Convolutional Neural Network (ES-FCN) model to autonomously segment the LV from the short-axis cardiac MRI data. To accommodate the 8-layer input of the ES-FCN, the input images are first transformed into 8 images using manipulators. Then, the altered images are put via a fuzzy-based intelligent edge detector to translate the greyscale into binary form, where the fuzzy module aids in handling the uncertain structural alterations in cardiac MRI. Then morphological operation fills the closed boundaries. Additionally, a new ES-FCN model is used to segment the LV part. This model was developed with a circulatory measure-oriented binary convolution layer, skip connection and bicubic interpolation. This research aims to address current challenges in short-axis cardiac MRI image segmentation by proposing a deep learning system for completely automated LV segmentation from a large dataset.

For this reason, a fully computerized method is needed to help clinicians perform more effectively. In this paper, we propose a combination of different deep neural networks for semantic LV segmentation based on Tri Convolutional Networks (ConvNets) to obtain highly accurate segmentation. Employing advanced deep learning models such as CNN (Sivasankari *et al.* 2022); LSTM (Fenil *et al.* 2019) and YOLO (Gayathri *et al.* 2023), and so on contribute to detecting different types of diseases.

1.11 OBJECTIVES OF THE THESIS

The chief and predominant objective of this research is to reveal the most suitable algorithm to investigate Heart Disease Detection using Deep Learning Techniques. The proposed algorithm is evaluated with the datasets and various algorithms are compared with the proposed algorithm and the accuracy is achieved with the dice index and Jaccard indices. Some of the specific objectives are elaborately given below.

1. This research work focuses on the Left Ventricle Segmentation of the Heart from the cardiac MRI datasets. Here instead of manual segmentation fully automatic segmentation is done.
2. The segmented images are applied to the proposed algorithm. Here Convolutional neural network techniques are applied for both the proposed algorithms.
3. Edge and Shape feature-based Fully Convolutional Neural Network (ES-FCN) algorithm is the first algorithm. Here it achieves the segmentation from the edge maps of the MRI image rather than using a gray-scale image.
4. The RoI-based Tri-Convolutional Neural Networks (Tri-ConvNets) algorithm is the second algorithm. Here it achieves the segmentation by giving the input as patches by extracting the noise-free images by using contour edge segmentation.
5. The proposed algorithms are made a comparative analysis with other algorithms by using the Datasets and it the accuracy is achieved.

1.12 ORGANIZATION OF THE THESIS

The thesis focuses on the Investigation of Left Ventricle Segmentation for Heart Diseases using Deep Learning Techniques, specifically Convolutional Neural Network (CNN) algorithms. In the First Phase, the Fully Convolutional

Neural Network concept is used to identify the Edge and Shape feature-based Fully Convolutional Neural Network for the Image datasets. In the Second Phase, Tri Convolutional Networks are used for the Image Datasets. Subsequently, both algorithms are compared with other algorithms and the achieved accuracy in the proposed algorithm is discussed along with the results.

The thesis is organized into seven chapters which are given as follows:

Chapter 1: General Introduction of the Thesis

In this chapter, the problem statement and solution, the motivation of the research work, imaging techniques, deep learning techniques, various convolutional neural network algorithms, and the organization of the thesis are presented.

Chapter 2: A Literature Review of the Existing Research Work

This chapter gives the literature review conducted for the research work is presented. The literature review includes a survey on Heart Disease, Left Ventricle Segmentation, and Convolutional Neural Network Algorithms.

Chapter 3: Fully Convolutional Neural Network Concepts

This chapter explains the Fully Convolutional Neural Network Algorithm concepts and the metrics used for evaluation.

Chapter 4: Automatic Left Ventricle Segmentation Using Edge and Shape Feature-based Fully Convolutional Neural Network

This chapter explains the Edge and Shape Feature-based Fully Convolutional Neural Network algorithm and its execution using the image datasets.

Chapter 5: Automatic Left Ventricle Segmentation Using RoI Based Tri-Convolutional Networks (ConvNets)

This chapter describes the Tri Convolutional Networks (ConvNets) algorithm and its working procedure using the datasets.

Chapter 6: Results and Discussion

The experimental results are discussed for both the proposed algorithm and it is evaluated.

Chapter 7: Conclusion and future scope of the research work

CHAPTER 2

LITERATURE SURVEY

Over the past ten years, there has been a lot of research focused on and attention to the Left Ventricle segmentation from the short-axis MR images. Examining functional heart measures such as ejection function, ES volume region, regional indexes, and wall mass is required for diagnosing cardiovascular disorders. Recently, there has been a lot of interest in the segmentation of the left ventricle using cardiac magnetic resonance imaging as a primary step toward this goal. Various artificially intelligent computer vision methods have been extensively used to pinpoint the exact location of the LV. Here, the existing LV segmentation techniques are examined, along with their significance and limits.

2.1 DEEP LEARNING METHODS FOR LV SEGMENTATION

Avendi *et al.* (2016) integrated the auto-encoders and CNN with a deformable model for LV segmentation. The limitation of this method is that it requires manual offline training and extensive hyperparameter tuning. Using twisted models and a deep learning method, a fully automatic LV segmentation device is calculated from short-axis cardiac MRI datasets. Deep learning methods are used in this procedure to extract the dissection task from the ground truth data. In the MRI dataset, convolutional networks continuously processing and interpreting the LV space. The LV outline is completed by applying loaded autoencoders. The dissection is made more accurate and robust by combining the conditional outline with twisted models. The method is proven to surpass state-of-the-art procedures and achieve an exceptional concordance with the ground truth. Forty-Five (45) cardiac MR datasets from the MICCAI 2009 LV dissection competition are used to validate it.

Vigneault *et al.* (2018) presented Ω-Net (Omega-Net): a novel Convolutional Neural Network (CNN) architecture for simultaneous localization, transformation into a canonical orientation, and semantic segmentation. Convolutional Neural Network (CNN) architecture allows localization, transformation into a canonical orientation, and semantic segmentation all at once. First, the input image is segmented; next, the features identified in the first segmentation are applied to predict the parameters needed to transform the input image into a canonical orientation, and finally, a final segmentation is applied to the converted image. They conclude that this architecture is a significant improvement over previous methods, with ramifications for biomedical picture segmentation in general.

Wu *et al.* (2020) combined the CNN and U-Net approach for cardiac segmentation, where CNN is used for locating the Area of Interest (ROI) and the U-Net for ventricular segmentation. To precisely segment the LV, they suggest using a composite model that combines CNN and U-net. So, the U-net network segments the LV after CNN locates the Area Of Interest (ROI). They have created a method known as modality adaption. They demonstrated the accuracy and robustness of the proposed model by training and testing it on cardiac MRI datasets from the MICCAI 2009 left ventricular dissection challenge. The measurements include Hausdorff Distance (HD), Volumetric Overlap Error (VOE), and Dice Metric (DM), with values of 3.641, 0.053, and 0.951, respectively.

Zhu *et al.* (2022) examined a large number of current articles to offer a comprehensive analysis of the diagnosis of breast cancer using the Convolutional Neural Network (CNN). The first step is to introduce a variety of imaging modalities and in the second part, CNN's organizational structure is described. Next, several open-access breast cancer data sets were presented. Then, a breast cancer diagnosis was carried into three distinct tasks: categorization, identification, and segmentation.

Xie *et al.* (2020) developed an autonomous intelligent model for segmenting the LV. In LV localization, deep CNNs were used in place of manual initialization of level set models. The MICCAI 2017 dataset was used for experimental evaluation, resulting in notable results. However, the established model only achieves localization of the LV under certain conditions, and if these conditions are not met, the results are negatively affected.

Graves *et al.* (2020) developed a deep-learning model with improved generalization capabilities to support images acquired in various environments. Histogram-based pre-processing method and traditional data augmentation technique were both used to enhance generalization. The U-net was then used to determine how well the pre-processing model worked.

Rostami *et al.* (2020) adopted a level-set technique and CNN-based LV. Here, the CNN model was used for the initial segmentation, and the semantic segmentation was supported by the level-set method. Intensity normalization was used to remove the unwanted artifacts from the image. It obtained a dice score of 0.94 when evaluated using the SCD dataset.

Zhu *et al.* (2023) designed a RBEBT model to automatically categorize brain tumors using RBEBT. A fine-tuned ResNet18 was used to extract features from photos of brain tumors. A Randomized Neural Network (RNN) was used for the RBEBT, which differs from conventional CNN models. RNN variables are currently being optimized using the Bat Algorithm (BA). Five-fold cross-validation demonstrates the superiority of the RBEBT by achieving 99.0% accuracy.

Amer *et al.* (2020) recommended dL framework ResDUnet, a U-net-based system with the dilated convolution united, is utilized. To address the issue of variation in LV dimensions and forms, incorporate a map of features produced by cascaded dilation into U-net's data synthesis procedure. The proposed model shows a dice similarity at 0.951% for LV segmentation.

Amer *et al.* (2021) proposed a fully computerized approach for LV segmentation that manages the LV margins and form variations while accurately delineating the ventricle boundaries. ResDUnet has been developed using the U-net architecture. By integrating cascaded dilated convolution, the characteristic extraction at various scales is identified. ResDUnet achieves a Dice consistency improvement of 0.95%.

Li *et al.* (2023) developed the multi-task neural network EchoEFNet to classify and segment LVs. The core used for gathering high-dimensional information while preserving spatial characteristics is the ResNet50 model. The biplane Simpson's approach was employed to seamlessly and precisely estimate the LVEF. The proposed technique was evaluated using the CAMUS and CMUEcho datasets, which yielded dice coefficient values of 0.936% and 0.936%, respectively.

2.2 AUTOMATIC LEFT VENTRICLE SEGMENTATION

Chen *et al.* (2020) give a review of deep learning-based cardiac image segmentation. The fundamentals of deep learning were covered, along with the different kinds of neural networks, such as fully convolutional neural networks, Generative Adversarial Networks (GAN), Autoencoders (AE), recurrent neural networks, and advanced building blocks for improved segmentation. Evaluation metrics, minimizing overfitting, and training neural networks using standard loss functions were all covered. Researchers looked at the cardiac MR image segmentation processes employing whole heart, aortic, ventricle, atrial, and scar segmentation. The availability of the dataset was also mentioned, along with the discussion of cardiac ultrasound and cardiac CT image segmentation.

Pluempitiwiriyawej *et al.* (2005) presented a level-set method for automatic cardiac image segmentation which hybrids the region, edge, and shape priors of the heart structure and local boundary properties. By reducing a liveliness purposeful, STACS associates local holdings of the curve and shape priors of the heart with stochastic region-based and edge-based information. The ability to segment images, the robustness to initial outline and noise due to the utilization of both edge and region-based information, and the ability to distinguish the heart from the chest wall and the unsought papillose physiques due to the existence of heart shape priors are the three most remarkable structures of STACS.

Kaus *et al.* (2004) presented a fully automated deformable model technique for cardiac segmentation of the LV in 3D cardiac MRI with mean error and end-diastolic phase. To segment a cardiac segment, standard hardware is needed. In cardiac imaging, the heart's position can be projected nearly to the picture's center. To extend our method's seizure range, we apply a coarse-to-fine variation methodology on an image resolution. The grey level variance and the annotations available in the cardiac images are adopted as prior knowledge.

Pednekar *et al.* (2006) presented an automated Left Ventricular (LV) myocardial boundary extraction using a motion map and an expectation maximization algorithm. An intensity-based fuzzy empathy map is then used to segment the myocardial border, and an energetic programming style is used to mine the myocardial outlines through cost reduction. It was discovered that the automated algorithm's results, when compared to those of the expert radiotherapists using Bland and Altman analysis, had a constant mean bias of 7% and limits of agreement equivalent to the inter-observer variability inherent in the manual method.

Hsu *et al.* (2019) presented a quicker active shape model and region-based convolutional neural network for automatically identifying, tracking, and segmenting the LV in cardiac illustration patterns. An enhanced adaptive anisotropic diffusion filter to efficiently strengthen image outlines and lower noise. An LV detection rate of 0.88% is achieved during the segmentation process. Hu *et al.* (2019) suggested a robust algorithm to increase the coronary MRI's automated LV segmentation speed. The key techniques developed in this segmentation algorithm are CNN coarse segmentation of LV images and ROI extraction. Using the SegNet method, the Jaccard indexes and epicardial contours are 0.80 and 0.76. The diverse sources of image variability can hinder analysis, assessing extreme slice images challenging due to unexpected variances.

Teixeira *et al.* (2008) proposed an active contour model based on snakes and a Genetic Algorithm (GA) to locate the LV boundaries in cardiac MRI images. The Snakes method has been modified specifically to autonomously segment the left ventricle of the heart in magnetic resonance imaging. To determine the ideal set of development restrictions for the Snakes approach, the genetic algorithm is employed. Tests are conducted using short-axis photos obtained from two reliable assistants. Based on initial findings, it appears that the methodologies are promising and could be utilized to explain the occurrence of the automatic cardiac ejection fraction scheming in the future.

Folkesson *et al.* (2008) present a segmentation technique that outspreads geodesic active region procedures by the assimilation of a statistical classifier trained using feature collection. The classifier gives class probability strategies built on class illustrative local features, and the geodesic active region generation allows the image to be segmented based on the area data. The automatic segmentation effects of the heart for cardiac late gadolinium-Enhanced Magnetic Resonance Imaging (CE-MRI) data utilizing coupled level set curve evolutions, where the classifier is integrated with the shape and the region term.

Ngo *et al.* (2013) introduced a new semi-automated methodology combining a level set method with Restricted Boltzmann Machine (RBM) for the problem of left ventricle segmentation, where the contour and edge facts are used as priors. One shortcoming here is that the method requires human interaction. The approach syndicates the level set advantages that use quite a lot of a priori truths about the entity to be segmented with the facts that are automatically learned from a manually marked-up database. Because deep trust networks can study robust models with few marked-up images and are flexible enough to be adjusted to a top-down separation challenge, their use is justified. The method produces mediocre results when used with the MICCAI Grand Challenge database for left ventricle segmentation from cardiac MRI images; in each of the challenge's processes, the policy products show a resemblance with the best in the field.

Petitjean *et al.* (2011) carried out a comprehensive study on methods for segmenting short-axis images using a cardiac cine MRI sequence that is both fully and partially automated. The study found that the current techniques had poor generalization, low accuracy, and low resilience. Nevertheless, without requiring user input, segmentation algorithms with strong prior knowledge are said to be able to greatly enhance the outcomes. Prior shape information must be included due to poor edge information and significant within-patient shape variance of the various sections. Because of the ventricle's complicated and varied form, as well as its thin and ill-defined margins that seem fuzzy in apical slices, this work remains crucial.

Tan *et al.* (2018) proposed a neural network regression classifier for automatic LV segmentation from both short- and long-axis scans. Automated cardiac segmentation from MRI datasets is essential for the prompt diagnosis and treatment of heart disorders. They recommend that the problem of automated left and right ventricular segmentation be addressed with a deep fully convolutional neural network design. Their model is efficiently taught from scratch in a single learning stage using whole-image inputs and ground facts, enabling it to conclude every pixel. This is the first that we know of, utilizing a fully convolutional neural network architecture for pixel-wise tagging in cardiac magnetic resonance imaging. Their model's robustness is proven by numerical studies, which show it to beat earlier fully automated approaches on a variety of cardiac datasets across different evaluation criteria. Furthermore, their model can quickly achieve state-of-the-art cardiac segmentation at a large scale by utilizing commodity computing resources like the graphics processing unit.

Sun *et al.* (2010) also presented deformable model-based segmentation, where the myocardial thickness and shape information are the priors. It is the best method to use median representations and to control the depth priors. Segmentation performance is estimated through a cross-validation policy. The automatic segmentation method is quite exact relative to manual segmentation.

Zhuang *et al.* (2008) presented an image registration method with prior anatomical information for ventricle segmentation. A normal heart diagram and a cadaver heart diagram were cast for the tests to record in-vivo data for ventricular segmentation proliferations. The findings have demonstrated that the segmentations based on the approach are reliable, accurate, and visually sufficient for in-vivo cardiac magnetic resonance imaging. The tests further show that the new initialization technique can improve the segmentation precision quantitatively while accurately adjusting for local misalignments and preventing idealistic distortions in nonrigid registrations.

2.3 MAGNETIC RESONANCE IMAGE SEGMENTATION

Duan *et al.* (2019) discussed about a multi-tasking deep learning approach with diagrammatic representation to develop a shape-refined bi-ventricular segmentation pipeline for short-axis CMR volumetric images. Thus, the methods for developing efficient segmentation of left ventricular from cardiac images are diverse. 3D MR images with high resolution are used for diagnosis; however, the automated systems with higher resolution are computationally expensive. An explicit shape of the preceding information, thus consenting exact, smooth, and structurally expressive bi-ventricular separations despite objects in the heart volumes. This research work focuses on LV segmentation with 2D slices as it has been reported as more efficient than 3D images.

Robinson *et al.* (2017) presented a segmentation method while considering the misclassification measures, and demonstrated that the approach can discriminate the successful and failed cases efficiently. The trend toward extensive research including human imaging presents new challenges for Quality Control (QC). When automated tools like image dissection techniques are used to extract biomarkers or quantitative measurements for further analysis, this becomes a complex problem. On a broad scale, manual inspection and graphic quality control of every dissection result are impractical. However, it's important to have the ability to recognize when an instinctive approach falls short of preventing inaccurate measures from appearing in subsequent studies, which could otherwise result in incorrect judgments. To solve this difficulty, we investigate a method that allows us to distinguish between successful and unsuccessful segmentation quality predictions based on reverse arrangement and accuracy. We approved this technique on a large cardiac MRI unit for which manual quality control scores were available.

Nambakhsh *et al.* (2013) proposed a distribution matching-based segmentation, where distance-based shape prior and intensity priors are used as matching constraints. This work examines the rapid detection of the epicardium and endo-LV shells in cardiac Magnetic Resonance Imaging (MRI) by using convex relaxation and circulation. For the algorithm to function, it requires a single subject and relatively little human input, totaling one point for each goal section in each cavity. By improving two functionals, each comprising two circulation matching limits such as a distance-based outline preceding and a strength preceding, it searches out each cavity section within each 3D stage. The shape above was created based on a universal amount of similarity between circles, making it intrinsically invariant to variation and change. To establish scale-invariance with simply a Fixed-Point Equation (FPE), a scale variable is introduced. This approach, in contrast to model-constructed priors, can withstand shape distortions and eliminates the requirement for expensive position valuation metrics and extensive training sets. For each of the real-world ones, the issue is broken down into a set of smaller issues that can all be addressed exactly and jointly with the help of an enhanced Lagrangian technique and a convex relaxation. The projected convex-relaxation resolution, in contrast to the associated graph-cut approaches, has no grid-bias problem, spreads out linearly to a substantial extent, and may be parallelized to significantly reduce the computational time for 3D fields. Parallelized enactment on a Graphics Processing Unit (GPU) indicates that a standard cardiac MRI volume takes 3.87 seconds to complete. This represents a speed increase of around five times compared to an archetypal execution. Additionally, they demonstrate that the process's performance is not expressively impacted by the theme that works best, and the outline explanation prevents expressively shifting from one topic to another.

These results demonstrate the situation where a single topic suffices to solve the provided algorithm.

Woo *et al.* (2014) presented a level set method with shape prior for LV segmentation. Heart disease analysis and quantitative estimation have continued to make extensive use of cardiac Magnetic Resonance Imaging (MRI). Cardiac MRI methods have remained increasingly enhanced, providing high-resolution functional and practical data. One of the important stages in the valuation of heart disease is the quantifiable examination of the Left Ventricle (LV) contractile purpose. Thus, the exact description of the LV frontier is of boundless attention to progress analytical presentation. So, an innovative dissection procedure of LV from cardiac MRI includes an implied shape preceding short of any working out stage through equal groups in a variational outline. The dissection of LV consists of motionless fragments and an interesting problem owing to its delicate borderline, obstruction, and inhomogeneity. To overwhelm such complications, an outline preceding information on the functional constriction of LV is combined into a region-based dissection context. The outline preceding is acquainted and created on the structural outline comparison among endocardium and epicardium. The outline of the endocardium is supposed to be commonly lower than mounting to the outline of the epicardium. An implied outline depiction using employed distance function is acquainted with and their inconsistency is restrained in a prospect way. Our outline constriction is executed by a common resemblance of outlines short of any working out stage that needs a gathering of outlines to acquire their numerical assets. Thus, the presentation of this technique has been proved on fifteen clinical datasets, viewing its probability as the foundation in the clinical analysis of heart disease.

Hadhoud *et al.* (2012) presented a kNN-based classification with shape inference for automatic ventricle segmentation. The LV endocardium and epicardium may be extracted from MR images using an improved ResUnet.

Enabling users to get used to using a medium skip connection for the narrowing path and a short skip connection for the remaining unit, increases the accuracy of the standard. Furthermore, to improve training efficiency, a depth-wise detachable involvedness modifies the standard convolution process. Using the MICCAI 2009 LV segmentation challenge test dataset, the average perpendicular distance of the endocardium, the dice measure, and the percentages of "good" contours are calculated. The results show that the model achieves promising performance and outperforms the advanced approaches. As a result, the model's segmentation accuracy is suggestively increased, and its efficiency is further advanced by the depth-wise distinguishable convolution.

Margeta *et al.* (2011) proposed a spatio-temporal forests-based machine learning approach for LV segmentation from 4D cardiac data. By clearly agreeing with the segmentation guidelines, the dataset is employed with a machine learning approach that uses two layers of spatiotemporal decision forests with almost no assumptions on the input. A method for context-aware MR intensity standardization and image configuration was devised, and the second layer was then eliminated for the final image segmentation. The 4D spatiotemporal characteristics are then added to the decision forests organization.

Mahapatra *et al.* (2013) presented a graph-cut framework for cardiac segmentation, which incorporates prior shape information. In this work, cine cardiac magnetic resonance images from the STACOM database are used to segment the myocardium, left ventricle, and right ventricle. The shape prior is favored for each dataset since it allows for a clear view of the many components in its image. A combination of distance goals and the placement of angle histograms for every pixel about the previous shape yields the previous shape information. Using only intensity data, which serves as the foundation for the second phase, segmentation is finished in the first phase. The final segmentation is obtained by combining intensity and shape statistics. The experimental results demonstrate advanced segmentation accuracy utilizing shape information on various subsets of thirty genuine patient datasets.

Santiago *et al.* (2017) provided a framework to deal with the variable number of slices in the training set and a resampling strategy for the test phase to segment the LV with any number of slices. A set of annotated volumes is used by three-dimensional active shape models to learn a shape model. By employing different landmarks to construct surface models in the training set, the shape model learns the anticipated form and variation modes of the segmentation. The segmentation procedure then proceeds with the application of shape constraints based on this information. If these models are used, then segmenting the left ventricle in 3D MRI volumes is an important problem. This challenge's annotations are related to a set of contours that make up the LV boundary for each volume slice. However learning the model is difficult because every volume has a different number of slices and, hence, a different number of landmarks. Moreover, the wide intervals between slices and motion artifacts.

Santiago *et al.* (2018) proposed a model to analyze and diagnose heart function, cardiac magnetic resonance imaging requires the segmentation of the Left Ventricle (LV). Cardiologists still have to manually complete this step in the majority of clinical setups, which takes time and is difficult. So, they study a quick approach for LV segmentation that greatly minimizes human intervention. The LV boundary is obtained with the use of a dynamic programming technique. This approach can handle many of the difficulties related to this topic by making very basic assumptions about the predicted position and shape of the segmentation. The computational complexity and accuracy of segmentation were assessed quantitatively. A left ventricle segmentation system for cardiac magnetic resonance imaging is suggested. Cardiologists no longer need to spend as much time or effort on their work thanks to this quick foundation. Later, they planned a faster segmentation approach based on dynamic programming using shape and location as priors.

Hu *et al.* (2019) presented a hybridized deep learning and dynamic programming framework for automatic segmentation of the LV from cardiac MRI datasets. Their suggestion is to utilize an elliptical active disc for automatic left ventricle segmentation in MR images, together with a multiscale template matching technique for detection. The local energy function is optimized by the elliptical active disc about the five free parameters that constitute the disk. Green's theorem is utilized in conjunction with gradient descent to minimize the energy function and optimize computing costs. 320 scans with 5,273 annotated slices that are made publicly available through the Multi-Disease, Multi-Centre, and Multi-Vendor Cardiac Segmentation (M & Ms) Challenge were among the findings they validated. The left ventricle was correctly located in 89.63% of the cases, and Dice coefficients of 0.873 and 0.770 were obtained for diastole and systole slices, respectively. The suggested method's performance on conventional image processing techniques is based on deep learning techniques.

Cong *et al.* (2019) proposed a Modality Adaptation Shape regression method (MA-Shape) for LV segmentation on mixed-modality images. This module effectively improved the performance of LV shape regression and the cross-modality generalization of MA-shape by extracting common and intrinsic shape features in each modality. Recently, they created a method for LV segmentation from mixed-modality images dubbed Modality Adaptation Shape regression (MA-Shape). In contrast to single-modality or multi-modality methods, they proposed that MA-Shape can be applied to images of new modalities during the test phase, which improves the generalization of the learned methods and makes use of already-existing samples of various modalities, thereby reducing the high demand for multi-modality data. Therefore, they provide a modality adaptation module to enhance the shape consistency between the

MR and CT to promote the generalization of the model learned in one modality to new modalities. These benefits allow for flexible and efficient use of automated heart function evaluation in clinical practice, in addition to offering an efficient means of leveraging mixed-modalities data during model learning.

Budai *et al*. (2020) proposed a regression-based solution. However, the regression-based methods are sensitive to rotation and translation variance. Regression-based fully automated technique for left and right ventricle segmentation is given. Their dataset included 5570 MRI short-axis scans of people who had CMR assessments for assessment and training at the Heart and Vascular Center, Semmelweis University Budapest. They have devised a novel method, and after training their algorithm on their dataset, it beat the state-of-the-art approach on both ventricles. Additionally, they compared the system's performance to human-level segmentation on both ventricles and discovered that the method performs similarly to expert and experienced readers in the left ventricle and readers in the right ventricle, respectively. They also tested the suggested technique with and without transfer learning on the publicly accessible ACDC dataset. ACDC results were similarly gratifying to human observers.

Grosgeorge *et al*. (2013) also suggested a graph cut-based segmentation with shape prior in which the form variations are computed using PCA. Here, the forms are represented by distance maps of their contour rather than a collection of dots. Due to its significant shape irregularity and poor contrast with surrounding materials, segmenting the RV is a challenging task. The forms are represented by distance maps of their contour rather than a set of dots, which reduces the requirement for an inflated breakthrough recognition and accompanying progression. Consequently, a shape model is produced via a PCA on the shape variations. This prior is superimposed over the image with very little user interaction to help with segmentation, and then integrated with the well-known graph cut architecture. A dataset that is available to the public has 248 MR images segmented using this technique.

2.4 FULLY CONVOLUTIONAL NETWORK METHODS

Bai *et al.* (2017) proposed an FCN-based semi-supervised learning approach, in which a segmentation network is trained from both labeled and unlabelled data to reduce the training samples and time.Deep learning-based Semi-Supervised Learning (SSL) algorithms have shown promise in medical imaging and can improve doctors' exclusive explanations by using unlabeled data. Nevertheless, most existing SSL algorithms in fiction tend to tamper with networks and/or data to standardize the traditional solution. recognizing a large number of data stages when teaching multi- or dual-tasking seems to have inherent estimate concerns. In essence, we use a two-task deep network that predicts both a geometry-aware near-conventional representation of the objective and a pixel-wise dissection technique together. Through a unique duty renovate layer, the nearly conventional picture is transformed into a dissection map that resembles one. At the same time, they present a dual-task reliability regulation among the near conventional derivative dissection plans and directly projected dissection plans for both categorized and uncategorized facts. Widespread researches on two communal datasets display that our process can mostly progress the presentation by integrating the uncategorized facts.

Poudel *et al.* (2016) proposed a Recurrent Fully-Convolutional Network (RFCN) that is trained end-to-end and combines anatomical identification and segmentation into a single architecture. By significantly reducing computation time, this can streamline the segmentation process and even enable real-time applications. It is feasible to occupy particular organizational and functional dimensions thanks to a fully automatic segmentation of heart magnetic resonance imaging. Consequently, a Recurrent Fully-Convolutional Network (RFCN) is employed, which may regulate inter-slice spatial dependencies over internal remembrance elements and absorb picture descriptions after the filled mass of 2D slices by combining functional recognition and dissection into a single, end-to-end skilled construction, RFCN can reduce computation time, create a dissection channel, and potentially enable real-time applications. The publicly available MICCAI 2009 Challenge dataset is included in the statement for the analysis of RFCN utilizing two datasets. Assessments between deep limited Boltzmann machines and fully convolutional networks, as well as a common type that drives inter-slice spatial connection, have been reached.

Khened *et al.* (2018) presented an enhanced up-sampling procedure for FCN-based MR image segmentation and reported a significant improvement in the performance measures. Medical picture segmentation has demonstrated significant potential for designs based on deep Fully Convolutional Neural networks (FCNs). These architectures, however, typically include millions of parameters and insufficient training samples, which results in over-fitting and subpar generalization. So, they introduced an innovative FCN-based architecture for medical image processing that is very parameter and memory-efficient. For FCN-like architectures, provide a unique up-sampling approach that avoids the feature map explosion by combining long skip and short-cut links. So, they suggest incorporating the parallel structures found in the Inception module to process the incoming images at numerous scales and viewpoints simultaneously. Additionally, they provide a unique dual loss function whose weighting scheme combines the benefits of dice loss and cross-entropy. Using two publicly available datasets, the Automated Cardiac Disease Diagnosis Challenge (ACDC-2017) and the Left Ventricular Segmentation Challenge (LV-2011), they have verified the suggested network design. Their approach is combined with cardiac segmentation and illness diagnostics to provide a fully automated, computationally efficient framework that can be applied in different scenarios.

Bai *et al.* (2018) proposed a pixel-classification method on a large dataset and reported dice indices over 0.90. In picture pattern identification and segmentation for a wide range of applications, deep neural networks have demonstrated considerable promise. Here, they provide an automated method based on a Fully Convolutional Network (FCN) for CMR picture analysis. A massive dataset from the UK Biobank, comprising 93,500 pixel-wise labeled photos and 4,875 subjects, is used to train and assess the network. Numerous technical metrics, such as the Hausdorff distance, mean contour distance, and Dice metric, have been used to assess the method's performance. Furthermore, clinically significant measures have been employed, including mass (LVM), End-Diastolic Volume (LVEDV) and End-Systolic Volume (LVESV) of the Left Ventricle (LV), and End-Diastolic Volume (RVEDV) and End-Systolic Volume (RVESV) of the Right Ventricle (RV). They demonstrate that when it comes to evaluating CMR images and determining clinically meaningful metrics, an automated approach performs on par with human specialists.

Chen *et al.* (2020) presented a comprehensive study on Deep Learning for cardiac image segmentation and reported the challenges as scarcity of labels, model generalization across domains, and interpretability. Also suggested the future directions that the requirement of smart imaging, data harmonization, and privacy. The proposed ES-FCN method uses a simple architecture that involves binary operations and uses shape feature descriptions with lower resource requirements, contributing to a better learning procedure and greater segmentation accuracy.

Abdeltawab *et al.* (2020) created an autonomous LV segmentation methodology with the FCN1 model, the LV region was initially extracted. Additionally, the nodules in the LV are more accurately quantified by the bottleneck layer fused FCN2 network. A deep learning method for the automatic measurement and segmentation of the left ventricle from cardiac cine magnetic resonance data. FCN design is used for the LV blood pool center point's dependable localization. The LV cavity and myocardium are then segmented using an ROI. The framework was validated using the ACDC-2017 dataset, yielding improved recognition and precise estimates of cardiac factors. The proposed model shows a dice similarity at 98.13% for the LV segmentation.

Wang *et al.* (2020) developed a novel FCNN to segment the LV without missing any local features. To get all attributes in the resultant image adaptive pixel-wise weighting method is integrated with the FCNN model. The efficiency of the established strategy was carried out with the CAP database. However, this model is under training due to the small number of dataset images.

Lin *et al.* (2020) developed a data augmentation approach to train FCN quickly in small samples using LV segmentation. It is possible to solve the regular overfitting problem of neural networks with this new shape model-based augmentation approach. An experimental analysis revealed that a limited number of samples sufficiently trained the new strategy.

2.5 EDGE SHAPED FCN METHODS

Hu *et al.* (2014) proposed an automatic LV segmentation based on Local Binary Fitting (LBF) model and dynamic programming techniques. After magnetic resonance imaging, left ventricular segmentation is crucial to determining both global and local cardiac function. Consequently, a methodology for segmenting the left ventricle on short-axis cardiac Magnetic Resonance Imaging (MRI) was established to advance the presentation of Computer-Aided Diagnostic (CAD) systems. Therefore, an automatic dissection technique for the left ventricle is contained in the source of the Local Binary Fitting (LBF) model and energetic software design approaches. 45 case data sets are used in the authentication research projects. The average upright distance is nearly 2 mm, and the percentage of acceptable outlines for both the endo- and epicardial outlines of our results is roughly 93.5%. Nearly 0.91 is the overlying dice metric. The fortitude coefficient, reversal, and our expected procedure on the LV mass good among the authorities. The Ejection Fraction (EF) is 1.076. As a result, the intended dissection technique presents information more effectively and has a high chance of increasing the precision of computer-aided analysis systems in cardiac disorders.

Pham *et al.* (2016) presented a shape-based active contour model for LV segmentation. This article examines a unique energy curve model for cardiac magnetic Resonance Imaging (MR) left ventricular segmentation. The intended dynamism practical, which has an information term and an outline term, joins the previous outline with strength data. To manage the signal of the early curve to desired entity borders, the information term is cast off from a region-based energetic outline model. Meanwhile, the shape term is used to constrain the expanding outline of the positioned form, which helps the archetype contract with imagery when there is object obstruction and contextual chaos. They only use the Kernel Principal Component Analysis (KPCA) to update the previous shape, which enables the previously acquired form to match the expected object's contour exactly. Additionally, they maintain the shape standardization process rather than solving a set of Euler-Lagrange equations as in conformist approaches, which helps to emphasize the variations between the outlines.

Patravali *et al.* (2017) proposed an end-to-end Deep Convolutional Neural Network (CNN) for MR image segmentation. To achieve the best model performance, this method is assessed using the ACDC Dataset, and several important insights are explored. Here, a 2D and 3D segmentation channel using Deep Convolutional Neural Networks (CNN) is created for fully automated cardiac MR image segmentation. The ACD Challenge 2017 dataset, which consists of 100 studies, is used to qualify these models end-to-end from scraping. Each study includes cardiac MR images in the End Diastole and End Systole stages. As a result, in terms of detachment metrics, both dissection models achieve close to cutting-edge presentation cuts, and they both demonstrate irrefutable accuracy in terms of clinical restrictions. By familiarizing oneself with a novel dice loss function and its classification with cross-information loss, a proportionate analysis is offered. By discovering diverse system constructions and comprehensive experiments, we converse numerous significant perceptions to gain an ideal model presentation, which is also essential.

Wu *et al.* (2021) developed an adversarial learning framework to autonomously segment the LV. A pose estimation model with dilated convolution is used to retrieve the rich-level features. The performance model was evaluated using the publicly accessible MICCAI dataset. The great density of the structures increases the training complexity of the network.

Cui *et al.* (2021) developed an attention-guided U-Net network was formed to segment the LV. This model included an attention model to suppress irrelevant features while focusing on domain-relevant features. Furthermore, an FTL loss function was introduced to tackle the data imbalance challenge. The LVSC dataset was used to test the effectiveness of the proposed model, and it obtained a Jaccard index of 0.75.

Shi *et al.* (2021) constructed a segmentation model utilizing the level-set approach to avoid the complicated training process of the classic FCN algorithm. The convexity of the method was regulated by level-set contours. The results of the experiments suggest that this technique is appropriate when there are inadequate benchmark images.

Wang *et al.* (2022) established an automatic LV localization model by utilizing the advantages of deep neural networks. By fusing the multi-scale characteristics and the dilated convolution different perceptual semantic information was obtained. In the decoder module, a full-size skip connection assists in fully utilizing the collected features.

The deep learning-based segmentation methods perform the training with binary masks and model the segmentation problem as a pixel classification problem, where each pixel should be labeled to their corresponding ventricles (Tran 2016; Jang *et al.* 2017; Panichaud *et al.* 2019; Vesal *et al.* 2019 and Sander *et al.* 2019). For instance, Tran (2016) developed an FCN for locating the ventricular contours.

2.6 TRI CONVOLUTIONAL METHODS

Yang *et al.* (2020) recommended using RetinaNet for Multiview echocardiography to recognize A2C, A3C, and A4C visuals and locate the LV. The LV recognition efficiency is measured and the mIOU metrics for A2C, A3C, and A4C are 0.858, 0.794, and 0.838, correspondingly. The Feature Pyramid Network (FPN), is utilized for the classification task.

Zhu *et al.* (2023) developed a ReRNet based on ResNet50 and it is the foundational model for feature extraction. The retrieved features are fed into three RNNs: the Schmidt neural network, the extreme learning machine, and the DRVFL. ReRNet outputs are an ensemble of these 3 RNNs based on a majority voting process. Validation of the proposed network was performed using 5 * 5 cross-validations with an average accuracy of 99.97%.

Zhu *et al.* (2023) created a DLBCNet to classify blood cells. Blood cell-specific models (BCGAN) were developed to create synthetic pictures. This feature extractor uses a pre-trained ResNet50 as a backbone model. The extracted features are fed into the proposed ETRN to enhance the multi-classification performance of blood cells with an accuracy of 95.05%.

Leclerc *et al.* (2020) planned LU-Net to enhance 2D echocardiogram segmentation. The CAMUS dataset is used in the study, and the images are hand-chosen ROIs centered on the segmentation masks of guidance. The related images were trimmed from these ROIs to generate fresh data that were analyzed using the standard U-Net structure. Additionally, the technique analyses the left ventricular end-diastolic and end-systolic amounts, resulting in a mean correlation of 0.96 and a mean absolute error of 7.6 ml.

Irshad *et al.* (2023) introduced the Light U-Net model, a sophisticated histogram-based image-enhancing method for LV segmentation. Following improvement, the images are sent into U-Net's encoder-decoder architecture via a brand-new lightweight process paradigm. The MICCAI 2009 dataset yielded a dice factor value of 97.7% and was utilized as a validation tool for the suggested approach.

2.7 SUMMARY

The Left Ventricle segmentation from short-axis MR images has received much research focus and attention over the past decade. 3D MR images with high resolution are used for diagnosis and the automated systems with higher resolution are computationally expensive. Also the DL methods are a basic building block of many modern methods utilized in segment LV patching. It's less useful for real-time segmentation despite being expensive, difficult in terms of time, and requiring an extended training phase. This process and MRI imaging's limited capacity to adjust to different conditions are other hindrances. The proposed system focuses on LV segmentation with 2D slices as it is more efficient than 3D images. A novel ES-FCN model is developed using binary images as input, which is then processed using a binary convolution layer. In the proposed model greyscale images are converted to binary images to provide strong generalization capabilities. Also ES-FCN method uses a simple architecture which involves binary operations and uses shape feature descriptions with lower resource requirement, contributes to a better learning procedure and greater segmentation accuracy.

In the second phase a novel Tri-ConvNets approach is used to address the problems. So, a combination of different deep neural networks for semantic segmentation of left ventricle based on Tri Convolutional Networks (ConvNets) to obtain highly accurate segmentation. Here, the CMRI images are pre-processed to remove the noise artefacts and enhance the image quality. The RoI based extraction is carried out in three different stages to accurately identify the LV in noise free CMRI. These three ConvNets performs different operations to give different segmentation results and these results concatenate to calculate the circular similarity for attaining better LV segmentation output.

The Automated Cardiac Diagnosis Challenge (ACDC) dataset, and the LV Segmentation Challenge (LVSC) dataset are used for ES-FCN algorithm. The Sunnybrook (SB) Dataset, and the York dataset are used in Tri-ConvNets. From all the datasets the images are segmented into 2D slices and their sensitivity, specificity, Positive Predictive Value, Negative Predictive Value, Dice index and Jaccard index are evaluated and accuracy is calculated. Sensitivity measures the ratio of correctly classified segmented regions as LV. Specificity measures the ratio of negative classified segmented regions as LV. Dice and Jaccard indices are used to measure the overlap between the ground-truth and the predicted segmentation region.

Thus, in this chapter, various imaging techniques were studied. The deep learning techniques and the convolutional neural network algorithm were reviewed. The automatic Left Ventricle Image Segmentation was well reviewed and from the various algorithms studied the Edge Shape Fully Convolutional Network Algorithm and Tri Convolutional Algorithm were designed in the proposed algorithm for the image datasets studied in the reviewed literature.

CHAPTER 3

FULLY CONVOLUTIONAL NEURAL NETWORK

3.1 INTRODUCTION

A typical CNN framework consists of a convolution layer, nonlinear activation function, pooling layer, and fully connected layer as basic building blocks as shown in Figure 3.1.

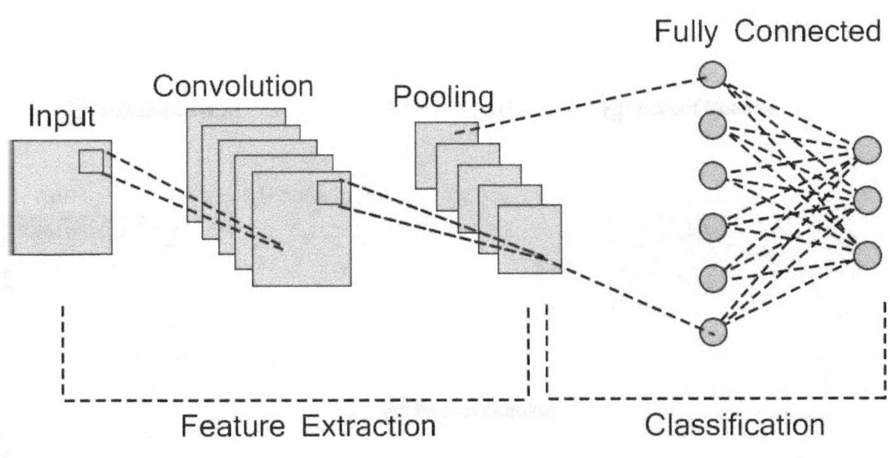

Figure 3.1 Baseline convolutional neural network architecture

The baseline CNN architecture results in the class probabilities for the image classification problem. However, the segmentation problem should involve pixel-wise classification rather than whole-image-based probability. (Long *et al.* 2015) addressed this issue and presented a CNN architecture where the fully connected layer is replaced with a fractional convolution layer as shown in Figures 3.2 and 3.3. Figure 3.2 illustrates an example of semantic segmentation with real-time images and Figure 3.3 depicts an example of medical image segmentation (Tran *et al.* 2016). The fractional convolution layer produces the output as pixel labels, which are validated against the ground truth segmentation, and the learning continued as similar to baseline CNN. The basic building blocks are explained in the following text followed by FCN discussion.

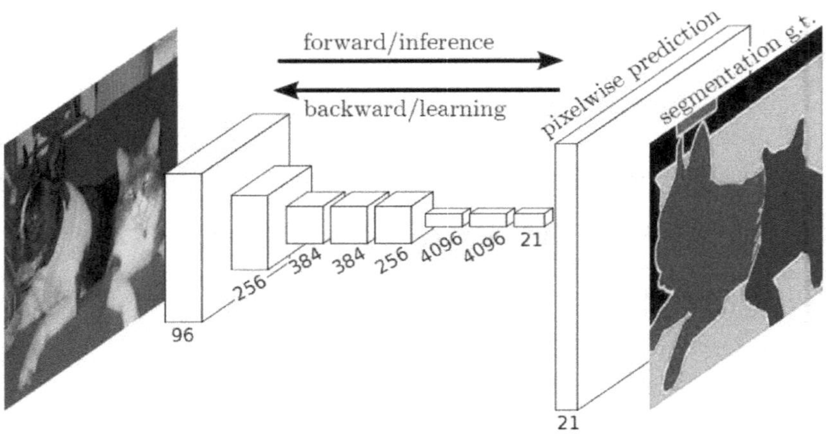

Figure 3.2 Fully convolutional networks for semantic segmentation

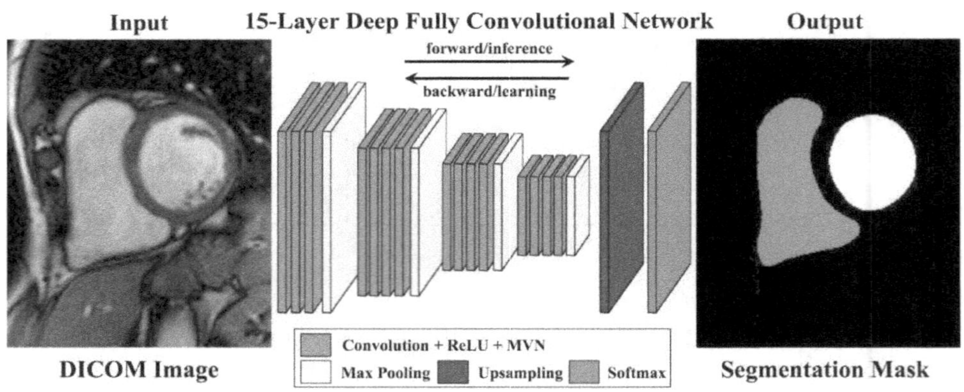

Figure 3.3 A schematic framework of a fully convolutional neural network for cardiac segmentation in short-axis MRI

CNNs are a category of neural networks simulated from the animal visual cortex (Gu *et al.* 2018). Hubel & Wiesel in 1959 studied the visual field of cats, identifying that visual cells respond to different types of stimuli. CNNs are designed to produce optimum performance from the translated, rotated, and distorted inputs. CNN builds local connections, and shared parameters, and makes use of several layers.

3.2 EVALUATION METRICS OF CNN

The convolutional layer is the primary operator of CNN, executed for the local image patch. The convolutional layer is composed of several kernels as shown in Figure 3.3, which are convoluted with the input to compute different feature maps. Every feature value is computed from a region of neighboring pixels (patch) from the previous layer. These kernel values (weights) are to be learned for better feature representation. To estimate each feature map, the kernel is shared among the entire image in a sliding window fashion. The shared kernels reduce the complexity of the classifier and improve the training performance. The complete feature set is computed by using several different kernels. The convoluted feature valuelocation (i, j) in the k-th feature map of l-th layer, $z^l_{i,j,k}$ is estimated as

$$z^l_{i,j,k} = w^l_k x^l_{i,j} + b^l_k \tag{3.1}$$

The Equation (3.1) mentions here w^l_k and b^l_k are the weight vector and bias values of the k-th kernel of the l-th layer respectively, and $x^l_{i,j}$ is the neighbourhood region centered at location (i, j) of the l-th layer.

The convoluted features are applied to an activation function to identify nonlinear features that are desirable for deep learning networks. The activation value $a^l_{i,j,k}$ of feature $z^l_{i,j,k}$ can be estimated as given in the below mentioned Equation (3.2).

$$a^l_{i,j,k} = f(z^l_{i,j,k}) \tag{3.2}$$

Sigmoid, tanh (Le Cun *et al.* 2012) and ReLU (Nair & Hinton *et al.* 2010) are the most commonly used activation functions. In general, the convolution and activation function operations are combined as one single layer. Rectified Linear Unit (ReLU) is one of the most noticeable non-linear activation functions and the same has been used in this research work, expressed as below mentioned Equation (3.3).

$$a^l_{i,j,k} = \max(z^l_{i,j,k}, 0) \tag{3.3}$$

The ReLU activation function can retain the positive feature value while pruning the negative part to zero. The simple max ( ) operation makes the ReLU operation faster when compare to the other activation functions like sigmoid and tanh. Also, ReLU prompts the network to maintain the sparse representation of the features. The literature shows that the deep learning architectures with ReLU function are achieving better generalization model though the discontinuity affect the routine of backpropagation (Russakovsky *et al.* 2015). And moreover, the issue of discontinuity at zero could be resolved by adding a constant parameter λ in the previous equation, known as Parametric Rectified Linear Unit (PReLU), defined as given in (He *et al.* 2015).

$$a^l_{i,j,k} = \max(z^l_{i,j,k}, 0) + \lambda_k \min(z^l_{i,j,k}, 0) \tag{3.4}$$

As mentioned in Equation 3.4 hereλ_k is the learned parameter for the kth kernel. He *et al.* (2015) shown that this extra parameter doesn't add much computational cost and there is no additional threat of overfitting too, moreover PReLU improves the accuracy.

Next, the nonlinear features are sub-sampled with a pooling layer to ensure the shift-invariance while decreasing the dimension of the feature maps. The pooling function ψ, for each feature map $a^l_{:,:,k}$ is computed as

$$p^l_{i,j,k} = \psi(a^l_{m,n,k}), \forall(m,n) \in \mathcal{R}_{i,j} \quad (3.5)$$

In the above given Equation (3.5) here $\mathcal{R}_{i,j}$ is a patch of neighbourhood pixels available around the location $(i,\ j)$. Max pooling and average pooling are the typical choices for pooling operations. Max pooling sample the patch with the maximum value, whereas the average pooling finds the average value of the patch. Recently, a combination of max and average pooling is proposed in (Yu *et al.* 2014), and the same has been used in this research work. The mixed pooling is given as mentioned in the below given Equation (3.6).

$$p^l_{i,j,k} = \beta \max_{(m,n)\in\mathcal{R}_{i,j}} (a^l_{m,n,k}) + (1-\beta)\frac{1}{|\mathcal{R}_{i,j}|}\sum_{(m,n)\in\mathcal{R}_{i,j}} a^l_{m,n,k} \quad (3.6)$$

where β is a random between 0 and 1, by choosing 0 for β, the mixed pooling would act as an average pooling function, the value 1 would turn this function as max pooling. The β value will be recorded during the forward process and the same will be used for the back propagation process. Yu *et al.* (2014) illustrated that this mixed pooling can resolve the over fitting issues and it outperforms max and average pooling.

The characteristics of kernels in CNN layers are designed in such a way that the kernels in earlier layers are meant for estimating low-level features such as edges and curves, while the later kernels are used to encode more abstract features. The higher-level features could be extracted by stacking more convolutional and pooling layers.

After several convolutional and pooling layers, a Fully-Connected (FC) layer is introduced to achieve high-level reasoning. The FC layer receives the inputs from the previous layer and connects them to every single neuron of the current layer to produce global semantic information. Also, the FC layer produces the classification result based on Softmax operator (Russakovsky *et al.* 2015). Let p represents the hyper-parameters of a CNN (e.g., the weight vectors and bias terms), they should be optimized by minimizing the classification error that could be modelled as a loss function. Consider a N-class classification problem, as given in the below mentioned Equation (3.7).

$$\{(x^{(n)}, y^{(n)}); n \in [1, \ldots, N]\} \tag{3.7}$$

where, $x^{(n)}$ is the n-th input data, $y^{(n)}$ is its corresponding target label and $o^{(n)}$ is the output of CNN. The CNN output of the j-th class for the i-th input is estimated as given in the below Equation (3.8).

$$o^i_j = \frac{e_{p^j}}{\sum_{t=1}^{N} e^{p^i_j}} \tag{3.8}$$

The softmax activation function transforms the output into positive values and normalizes them to receive a probability distribution over the classes. With M set of training samples, such outputs are utilized to estimate the softmax loss is mentioned in the below Equation (3.9) as

$$\mathcal{L}_{softmax} = -\frac{1}{N}\sum_{i=1}^{N}\sum_{j=1}^{M} 1\{y^{(i)} = j\} \log o^i_j] \tag{3.9}$$

The training phase is a global optimization problem, while minimizing the loss function; the hyper-parameters can be optimized for better classification accuracy. Stochastic Gradient Descend (SGD) is a common solution for optimizing CNN's hyperparameters (Wijnhoven *et al.* 2010 and Zinkevich *et al.* 2010).

A standard algorithm which utilizes the gradient descent approach to update CNN's hyperparameter is the backpropagation algorithm. The softmax loss functions is viewed as an objective function to optimize CNN's hyperparameter, $\mathcal{L}(\varphi)$, where the SGD method updates the parameters iteratively as

$$\varphi_{t+1} = \varphi_t - \eta \, \mathrm{E}[\mathcal{L}(\varphi_t)] \tag{3.10}$$

In the above Equation (3.10) here $E[\mathcal{L}(\varphi_t)]$ is the expectation of $\mathcal{L}(\varphi)$ with full training set and η is the learning rate. Instead of estimating E, SGD computes the gradients based on a single randomly chosen sample $(x^{(t)},\ y^{(t)})$ from the training set as given in the below Equation (3.11).

$$\varphi_{t+1} = \varphi - \eta_t\ \mathcal{L}(\varphi_t; x^{(t)}, y^{(t)}) \tag{3.11}$$

Computing the gradients based on single training sample might increases the variance in the parameter update, to address this issue, the gradients could be estimated based on mini-batch of training samples towards stable convergence. And the convergence speed could be controlled with an appropriate learning rate. Again, choosing an appropriate learning rate is another issue, one common solution is to start with a constant learning rate at initial stages, and then reduce it whenever the convergence rate slows down. To identify the velocity of the convergence rate, momentum is introduced to control the parameter update, is given as in the below mentioned Equation (3.12).

$$v_{t+1} = \gamma v_t - \eta_t\, \mathcal{L}(\varphi_t;\ x^{(t)}, y^{(t)}) \tag{3.12}$$

$$\varphi_{t+1} = \varphi_t + v_{t+1} \tag{3.13}$$

In the above mentioned Equation (3.13) here v_{t+1} is the current velocity vector gamma is the momentum factor, which is generally assigned with 0.9. A large number of parameters and non-convex loss function make CNN's training phase very complex (Choromanska *et al.* 2015). For a better convergence rate and to avoid the vanishing gradient issue, an appropriate kernel/weight and bias initialization is much required. For the baseline CNN architecture, the bias values could be set with zero, and the kernel should be carefully initialized to avoid symmetry among different layers. Constant or Random initializations are the popular solution, apart from that (Glorot & Bengio *et al.* 2010) proposed a Gaussian distribution-based weight initialization and reported significant CNN performance improvements.

3.3 RESEARCH METHODOLOGY

3.3.1 First Phase

The primary goal of this anticipated work is to propose a deep learning framework for fully automated LV segmentation from a huge dataset of short-axis cardiac MRI images in order to overcome existing roadblocks.

The major contribution of the proposed model in the First Phase is listed out below,

- A novel ES-FCN model is proposed using the binary image as input for segmenting the LV from the short-axis MRI image.
- Bicubic interpolation is used as an up-sampling operator to deconvolve the convolved images back to their original state.
- To convert the input MRI image into binary, fuzzy-based canny edge detection is exploited, and the fuzzy has the potential to handle the uncertainty scales, shapes, and structural changes of LV.
- To give a detailed view analysis to the physicians the input images are manipulated into 8 images and processed into the ES-FCN simultaneously.
- An LV segmentation framework based upon the fuzzy-based canny edge detection and ES-FCN model is introduced, and the efficiency of the model is evaluated by the publicly available datasets.

3.3.1.1 Experimental results of first phase

In the proposed Edge and Shape based FCN (ES-FCN) segmentation method it is evaluated with six performance metrics and compared with existing segmentation algorithms. The ACDC and LVSC datasets were used to evaluate the proposed segmentation methodology and produced Jaccard indices and it has superior and significant segmentation performance with higher Positive and Negative Predict values.

3.3.2 Second Phase

In the Second Phase we propose a combination of different deep neural networks for semantic LV segmentation based on Tri Convolutional Networks (ConvNets) to obtain highly accurate segmentation. By employing advanced deep learning models such as CNN, LSTM, YOLO and so on contribute to detect the heart diseases.

The major contribution of the proposed model in the second phase is as follows.

- Initially, the CMRI images are pre-processed to remove the sound artefacts and enhance the image quality. The RoI based extraction is carried out in three different stages to accurately identify the LV in noise free CMRI.
- The initial phase is to generate a bounding box on LV, then cover up the contour edges of LV with canny edge detector and finally the contour points are marked with the mask thresholding.
- Then, the extracted features are given as an input to the three different deep learning structures for segmenting the LV in the efficient way. Although a Transformer is utilized for extracting broad features from cardiac MR images, Tri-ConvNets gather fine-grained details and coarse-grained semantics from entire levels.
- Afterwards, the contour edges are processed in the standard ConvNet, the contour points are processed using Fully ConvNet and finally the noise free images are converted into patches to perform the pixel-wise operation in ConvNets.
- Finally, the three ConvNets performs different operations to give different segmentation results and these results concatenate to calculate the circular similarity for attaining better LV segmentation output.

3.3.2.1 Experimental results of second phase

The proposed Tri-ConvNets model used the Sunny Brook and York Datasets and it achieves the Jaccard indices and the dice index for the ACDC dataset and LVSC dataset.

3.4 SUMMARY

In this chapter the concepts of Fully Convolutional Neural network are explained. The evaluation metrics used for the Convolutional Neural network are explained in a detailed manner. The research methodology of first phase and second phase algorithms and their experimental results are elaborated in this chapter.

CHAPTER 4

AUTOMATIC LEFT VENTRICLE SEGMENTATION USING EDGE AND SHAPE FEATURE BASED FULLY CONVOLUTIONAL NETWORK

4.1 INTRODUCTION

A novel binary convolution layer-based ES-FCN model is introduced, which processes binary input images. The low-dimension binary images are used instead of greyscale images to reduce computation and complexity while improving the learning capacity of the model. To support the initial layer size of the ES-FCN, the collected images are first augmented into 8 images. Subsequently, a fuzzy-based canny edge detection technique is used for binary image conversion to address the structural changes that occur in the LV during blood pumping. There is a morphological procedure set up to fill the Ventricle and Atrium regions. The pre-processed images are taken as input to the ES-FCN which is made up of binary convolution layers, 2 x 2 max pooling, and bicubic interpolation-based deconvolution. Figure 4.1 provides the flow diagram of the proposed ES-FCN-based segmentation framework.

4.2 DATA PRE-PROCESSING

In general, image segmentation considers the grayscale image as input, proceeds with successive computation at various layers of CNN, and the kernel weights are learned based on the segmentation map. For such semantic segmentation, the kernel weights and other parameters are set with real numbers. For the proposed ES-FCN the computational cost involved in real number manipulation is reduced by considering the edges of the images as input rather than feeding the grayscale image. The images from the ACDC and LVSC dataset has been used to demonstrate the proposed framework. The proposed model comprises the following steps to translate the greyscale image into a binary image: (1) Data augmentation (2) Fuzzy based canny edge detection (3) Morphological operation

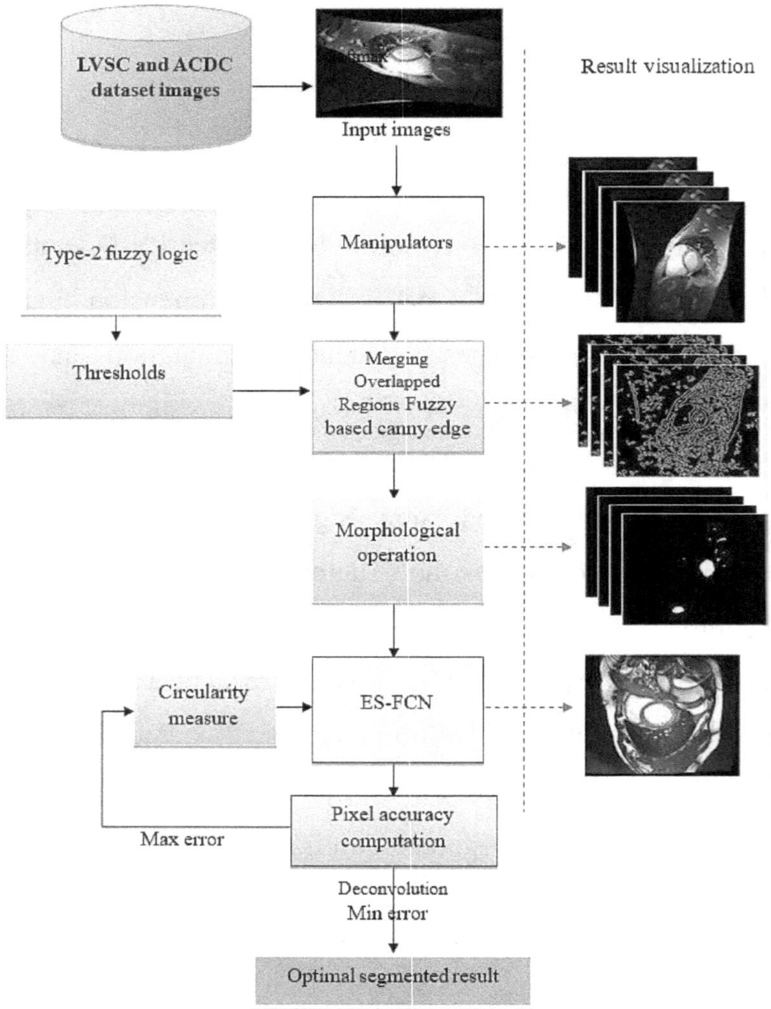

Figure 4.1 Architecture of ES-FCN-based segmentation model

4.2.1 Data Augmentation

The proposed ES-FCN features an 8-layer input, allowing for the application of various manipulators to convert a single image into 8 images.

Rotation: The image is rotated at a 45° angle.

Blurring the Image: The visual quality of the input image was diminished by applying a Box filter, Gaussian noise, Salt and Pepper noise, Gamma noise, and Uniform noise.

Smoothening: Median filtering and Wiener filtering are used to smooth the input image.

Once the MR image has been manipulated with the above-mentioned eight operators, the 8-layers of the input are ready to be fed as inputs to the proposed ES-FCN. The manipulation operation is depicted in Figure 4.2.

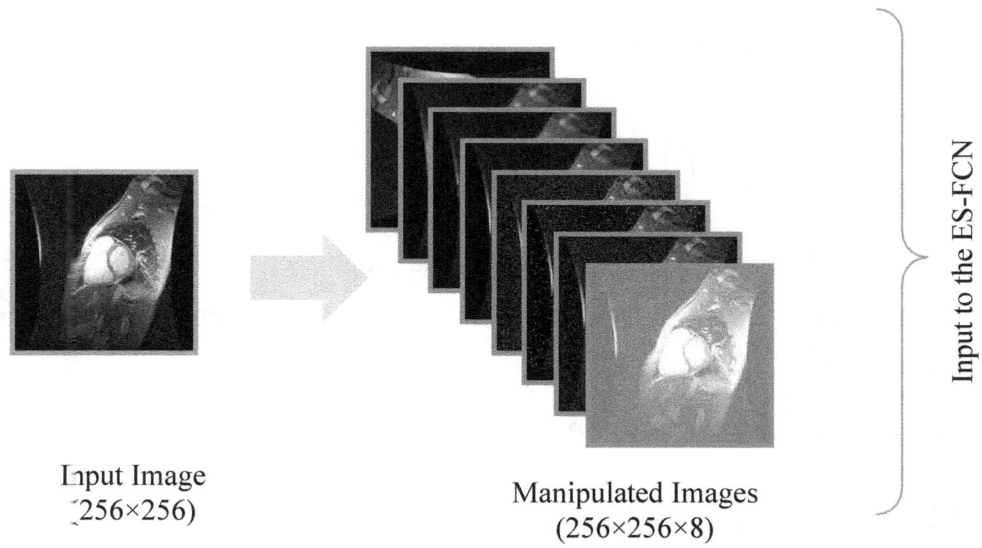

Figure 4.2 Manipulated input images for ES-CNN

4.2.2 Fuzzy-based Canny Edge Detector

The edges of the manipulated image are estimated with the 'Fuzzy-based Canny' edge detection operator. Although the traditional canny technique produces durable borders for the greyscale image, poor lighting and hazy region boundaries can introduce uncertainty into the gradient image. To handle the structural changes during the LV shrinks, a Fuzzy Sets-fused canny is exploited, by automatically selecting the threshold values it overcomes the uncertainties.

For the Canny operator, at first, the image is applied with a Gaussian filter to smoothen it and the Sobel operator is applied to highlight the regions by finding the gradients. Continuously, non-maximum suppression is applied to suppress the non-maximum pixels. Followed by a double thresholding model, choose the strong, weak, and non-relevant pixels. Finally, by employing hysteresis, the non-relevant pixels are processed. The following kernel, as given below in matrix 4.1 is used for finding the edges.

$$\frac{1}{115}\begin{bmatrix} 2 & 4 & 5 & 4 & 2 \\ 4 & 9 & 12 & 9 & 4 \\ 5 & 12 & 15 & 12 & 5 \\ 4 & 9 & 12 & 9 & 4 \\ 2 & 4 & 5 & 4 & 2 \end{bmatrix} \quad (4.1)$$

However, finding the threshold parameters $T1$ and $T2$ in the double thresholding procedure is challenging when the image boundaries are unclear. Therefore, by applying the Type-2 fuzzy set, the thresholds are dynamically updated based on the input data.

4.2.3 Threshold Selection with Fuzzy Logic

The single Threshold (T_h) selection method of Type-2 fuzzy set is used in the gradient histogram formulated by the Sobel operator to achieve the automatic threshold selection methodology. The gradient histogram is separated into two parts with the assistance of T_h, which then identifies the non-edge and edge points. The threshold is computed with the following Equation as given in (4.2).

$$UF = \frac{1}{M} \sum_{min}^{max} h(X)[\mu_U(X) - \mu_L(X)] \tag{4.2}$$

where UF is the ultra-fuzziness of the Type-2 fuzzy set, maximum of it returns the optimal threshold, N is the total number of data points in the histogram, and the X-coordinate maximum and minimum values are represented as max and min respectively. Point X histogram is depicted as $h(X)$. By shifting the lower $\mu_L(X)$ and upper $\mu_U(X)$ membership functions as in Figure 4.3, the UF value is computed. The hedge operator (β) is applied to construct the upper and lower membership functions at the base of the skeleton membership function $((\mu(X))$ as given below in the Equations (4.3) and (4.4).

$$\mu_U(X) = [\mu_U(X)]^{\frac{1}{\alpha}} \tag{4.3}$$

$$\mu_L(X) = [\mu_U(X)]^{\alpha} \tag{4.4}$$

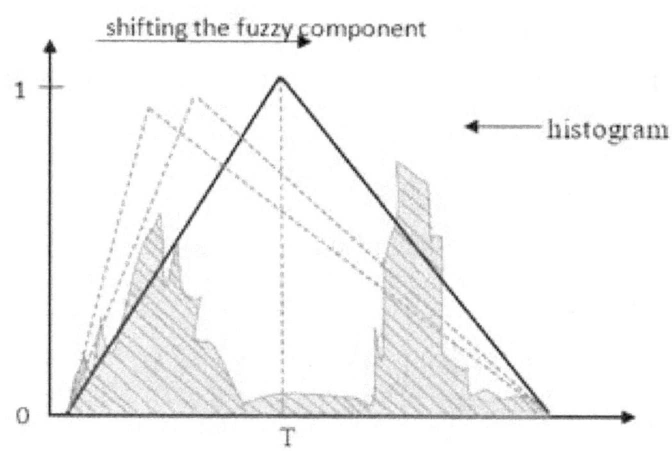

Figure 4.3 Shifting the membership function on the histogram

After detecting the threshold, the histogram (range 1 to range 2) is separated into two homogeneous modes, namely non-edge (range 1 to T_h) and edge ($T_h + 1$ to range2) portions. Further, the non-edge point is again divided into two parts by T_{h1} as non-edge points (range 1 to T_{h1}) and uncertainly non-edge points ($T_{h1} + 1\ to\ T_h$).

Simultaneously, the edge part is divided into uncertain edge points (T_h to T_{h2}) and edge part (T_{h2}+1 to range2). The formulated thresholds are further utilized in the canny operator to detect the edges. Then the uncertain portions ($T_{h1} - T_{h2}$) are processed in the hysteresis function. Due to the binary nature of edge images, the computation speed of FCN is faster when dealing with these images. This algorithm describes fuzzy-based canny edge detection.

Algorithm: Fuzzy-based canny descriptor for edge detection

Input: manipulated images I_i

Output: edge image E_i

begin

for each I_i**do**

Apply a gaussian filter to remove the noise

Compute gradients (range1 to range 2) with Sobel operator

Suppressing non-maximum pixels with non-maximum suppression

for array gradient histogram

Find T_h with type-2 fuzzy logic

Split the gradient into non-edge (range 1 - T_h) and edge(T_h+1 - range 2) points

if portion= range 1 - T_h

Split with T_{h1} into non-edge points (range1 to T_{h1}) and uncertainly non-edge points ($T_{h1} + 1\ to\ T_h$)

else

$T_{h1} = T_h$

---

end if

if portion= $T_h + 1$ - range2

Split with T_{h2} into uncertainly edge points (T_h to T_{h2}) and edge points (T_{h2}+1 to range2)

else

$$T_{h2} = T_h$$

end if

Apply $T_{h1} - T_{h2}$ into the hysteresis

end for

Apply the thresholds and find the edges

end for

4.2.4 Morphological Open Operation

A disk-shaped structuring element based simple morphological open operation is performed to fill the closed boundaries of the edge images and the corresponding results are depicted in Figure 4.4. It is expected that the Ventricle and Atrium regions would get filled by this morphological open operation.

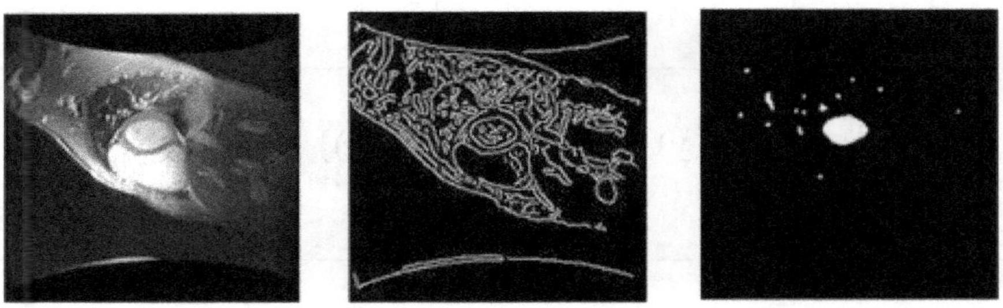

(a) (b) (c)

Figure 4.4 Illustration of morphological operation (a) Input image, (b) Edge image and (c) Morphological operation

4.3 EDGE AND SHAPE FEATURE BASED FULLY CONVOLUTIONAL NEURAL NETWORK (ES-FCN)

In general, image segmentation considers the grayscale image as input, proceeding with successive computation at various layers of CNN, and the kernel weights are learned based on segmentation map. For such semantic segmentation, the kernel weights and other parameters are set with real numbers. For the proposed Edge and Shape based Fully Convolutional Network (ES-FCN), the computational cost involved in real number manipulation is reduced by considering the edges of the images as input rather than feeding the grayscale image. In the first step, the short-axis MR image of LV is manipulated with eight different operators as listed in the following Table 4.1.

Table 4.1 Image manipulation operators applied over short-axis MR images

Operator	Parameter	Remarks
Rotation	90°	Image gets rotated with 90°
Box filter	$\frac{1}{25}\begin{bmatrix}1&1&1&1&1\\1&1&1&1&1\\1&1&1&1&1\\1&1&1&1&1\\1&1&1&1&1\end{bmatrix}$	Image blurring with a box filter of size 5×5
Median filter	$\hat{f}(x,y) = \underset{(r,c) \in S_{xy}}{median}\{g(r,c)\}$	Image smoothened with a median filter of size 7×7
Wiener Filter	$R(u,v) = \frac{H*(u,v)}{\lvert H(u,v)\rvert^2 + [\frac{S_n(u,v)}{S_I(u,v)}]}$	H – blurring filter, * - Complex conjugate $S_n(u,v)$ – power spectrum of noise $S_I(u,v)$ – power spectrum of image

Table 4.1 Continued

Gaussian Nosie	$p(g) = \frac{1}{\sqrt{2\pi\sigma}} e^{-\frac{(g-\mu)^2}{2\sigma^2}}$	g – gray value, μ – mean gray level, and σ – standard deviation
Salt & Pepper Noise	$p(g) = \{P_s \quad for\ g = 2^k - \quad P_p \quad for\ g = 0 \quad 1 - (P_s + P_p) \quad otherwise$	P_s, P_p – sale & pepper noise probabilities k - #bits per pixel
Gamma Noise	$p(g) = \{\frac{a^b z^{b-1}}{(b-1)!} e^{-ag} \quad g \geq 0 \quad 0 \quad g < 0$	a, b – are the random numbers
Uniform Noise	$p(g) = \{\frac{1}{b-a} \quad a \leq g \leq b \quad 0 \quad otherwise$	a, b – are the random numbers

The ES-FCN segments the LV using binary images with less complexity than the traditional LV segmentation method. Additionally, it provides adequate results in images derived from various environments due to its conversion of binary input images. The proposed model is constructed using the binary convolution operation biased with shape descriptors, maximum pooling, bicubic interpolation-based deconvolution, and softmax layers. Figure 4.5 depicts the proposed ES-FCN framework for LV segmentation.

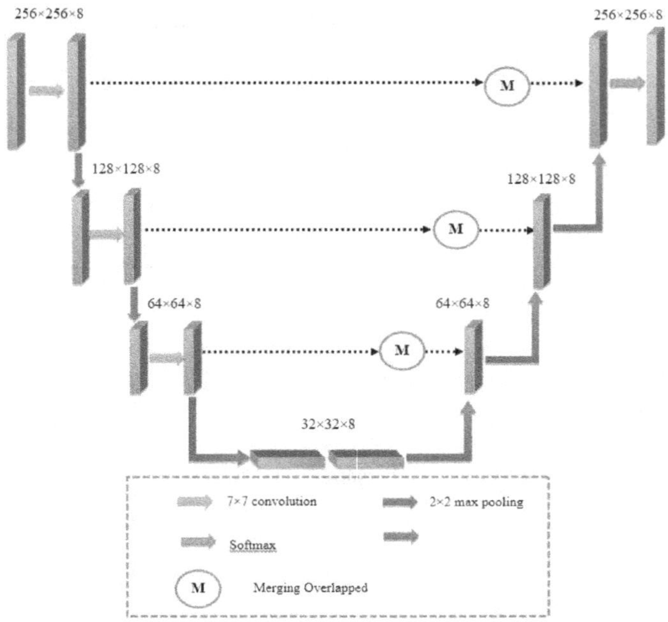

Figure 4.5 The proposed architecture of ES-FCN model for LV segmentation

4.3.1 Contracting Path Layers

For the convolution operation, a simple kernel with only 1 in the middle and keeping the other elements as 0, is used. For example, a kernel of size 3 × 3 is initialized as given in the Equation (4.5).

$$filter_{kernel_{3\times3}} = \begin{bmatrix} 0 & 0 & 0 \\ 0 & 1 & 0 \\ 0 & 0 & 0 \end{bmatrix} \quad (4.5)$$

Convolving images with such kernels would retain the same image without any alteration. The center element value will be assigned with either 0 or 1, based on the circularity measure of subimage. The circularity measures are computed as mentioned in the Equation (4.6).

$$C = \frac{(4\times Area\times\pi)}{Perimeter^2}, 0 < C \leq 1 \quad (4.6)$$

For every subimage to be convoluted, the Area and the Perimeter of the connected region from the edge map is measured, with $C$is estimated. From the dataset of short-axis MR images, it is understood that the LV region would be in a circular shape. Hence, the subimages having circularity measure greater or equal to 0.5 are retained by assigning 1 in the middle for the filter kernel, whereas for the other regions, the convolution filter will take all zeros, hence such regions are removed from the edge image. Figure 4.6 illustrates a sample output after a single step convolution operation. The center element of the filter kernel is either assigned with 0 or 1, based on circularity measure, given below in the Equation (4.7) as

$$w_k^l(r,c) = \begin{cases} 1 & C \geq 0.5 \\ 0 & otherwise \end{cases} \tag{4.7}$$

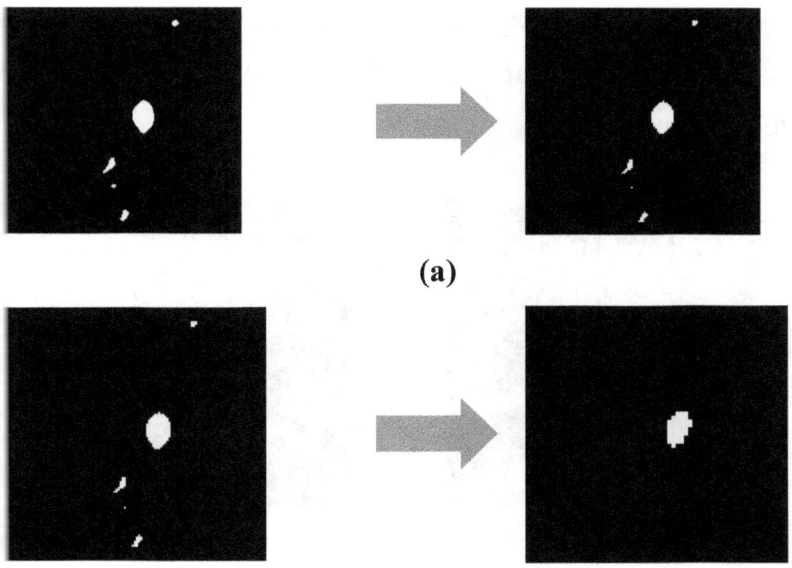

(b)

Figure 4.6 Sample convolution operation (a) with $C \geq 0.5$ and (b) $C < 0.5$

The input images are binary and the convolution operation is also a binary operation, hence the activation function operation is eliminated here. Consequently, the convolution operation is followed by the pooling operation. The conventional Max pooling operation is applied here with a smaller 2×2

window size; therefore, the resultant image size would be half of its original size. The convolution and max pooling operations are continued for another 2 levels at least. For an image of size 256×256, would be reduced to 128×128 at the first level, followed by 64×64 and 32×32 in the second and third levels, respectively.

4.3.2 Expansion Path Layers

The images are upsampled with bicubic interpolation and deconvolved the images back to their original size. At each step of deconvolution, skip connections are used to fuse coarse semantic data at deep layers and precise segmentation at the shallow layers to learn weights for output upsampling. The segmentation outcomes from the corresponding levels are utilized to improve the segmentation boundaries. For example, consider the given segmentation results I_{c1} and I_{D2} received after 1st level convolution and 2nd level deconvolution, as depicted in Figure 4.7.

(a) (b)

Figure 4.7 Convolution and deconvolution results (a) Convolution result after 1st Level and (b) Image after 2nd level deconvolution

Each bounded region of I_{c1} are compared with each bounded region of I_{D2}. Whichever the regions of I_{D2} falls exactly inside any of the regions from I_{c1} are merged with logical OR operation, while ignoring the other regions where there are no matching sub-regions available in I_{D1}. Figure 4.8 illustrates the skip architecture for finding the overlapped regions from the segmented results.

(a) (b) (c)

Figure 4.8 Refine segmentation by skipping connections (a) Bounded regions on convoluted result, (b) Bounded regions on deconvoluted result and (c) Overlapped regions are merged

The upsampled image received from the final layer is compared with the segmentation map received from the radiologists (o) and the predicted result (y). The prediction error is measured with Pixel Accuracy (PA), which defines the rate of precisely categorized pixels to the total number of pixels. Based on this error, the circularity measure is backpropagated and adjusted to train the model. The layers of images are processed at the sequence layers of FCN and the segmented output received at the final layer is evaluated by the ground-truth segmentation map.

4.4 SUMMARY

In this chapter, the architecture of the Edge and Shape Feature Based Fully Convolutional Network algorithm is explained. The pre processing of data, the morphological operation, the convolution and deconvolution of the images are explained in detailed manner.

CHAPTER 5

AUTOMATIC LEFT VENTRICLE SEGMENTATION USING ROI BASED TRI-CONVOLUTIONAL NEURAL NETWORKS

5.1 INTRODUCTION

A combination of different deep neural networks is utilized for the semantic segmentation of the left ventricle based on Tri Convolutional Networks (ConvNets) to obtain highly accurate segmentation. Initially, the CMRI images are pre-processed to remove the noise artefacts and enhance the image quality. The RoI based extraction is carried out in three different stages to accurately identify the LV in noise free CMRI. The first stage is to create a bounding box on LV, then cover up the contour edges of LV with canny edge detector and finally the contour points are marked with the mask thresholding. These features are given as an input to the three different deep learning structures for segmenting the LV in the efficient way. Tri-ConvNets have the ability to gather both smooth and coarse-grained semantic information from entire levels. The contour edges are processed in the standard ConvNet, the contour points are processed using Fully ConvNet and finally the noise free images are converted into patches to perform the pixel-wise operation in ConvNets. These three ConvNets performs different operations to give different segmentation results and these results concatenate to calculate the circular similarity for attaining better LV segmentation output. Figure.5.1. depicts the illustration of proposed Tri-ConvNets model.

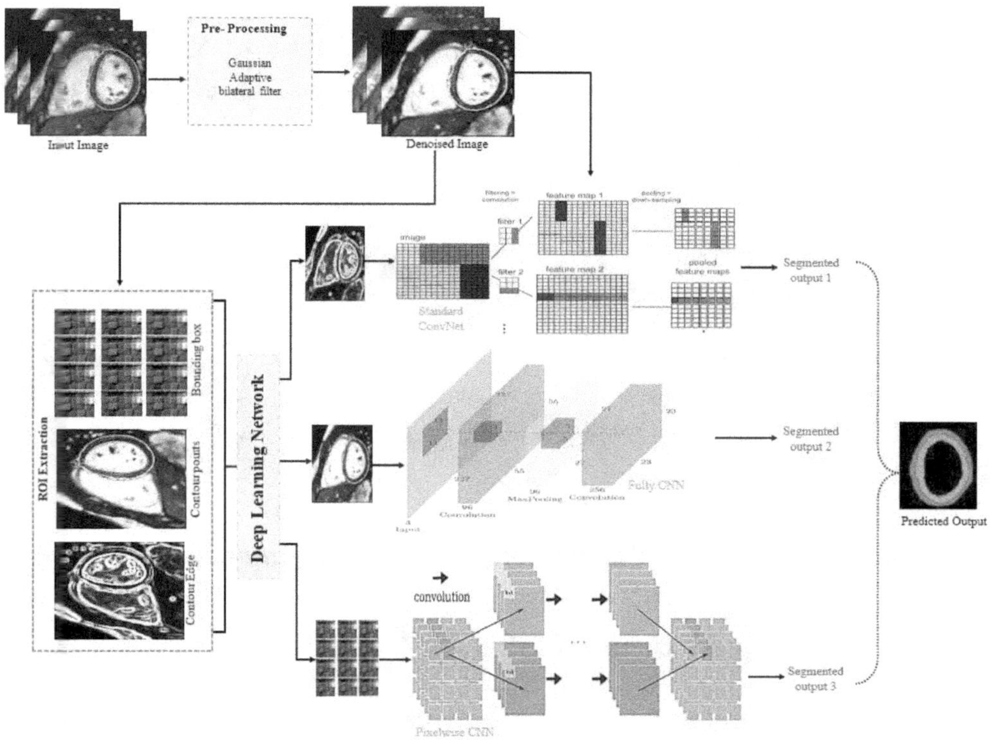

Figure 5.1 Schematic illustration of the proposed Tri-ConvNets model

5.2 ADAPTIVE BILATERAL HISTOGRAM EQUALIZATION FILTER (ABIHE)

Here an Adaptive Bilateral (AB) and histogram equalization filters are utilized to preprocess the input images to minimize noise distortion and improve image quality. HE is applied after the AB filter has smoothed the image to improve its quality and eliminate additional noise. The preprocessing of the left ventricle is exposed in Figure 5.2.

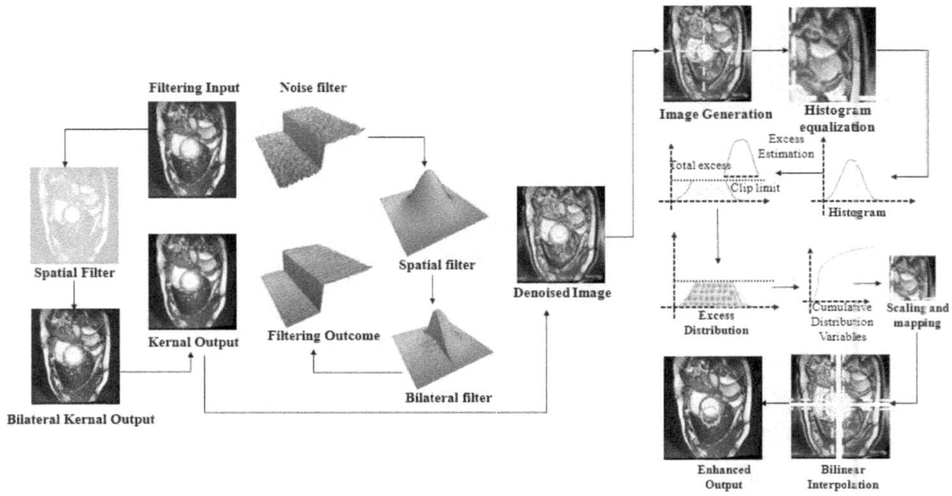

Figure 5.2 Schematic illustration of pre-processing process for left ventricle

5.2.1 Adaptive Bilateral Filter

The bilateral filter effectively minimizes noise in images while preserving edge features by selectively allowing pixels to contribute to the scaled average. Its application results in smoother images, and various filters within this category can also identify object boundaries in images. Smooth images with preserved essential data are achieved through the use of the bilateral filter, as illustrated in Figure 5.2. The bilateral filter yields optimal results, and pixels on one side of the boundary influence the altered pixel value due to bilateral effects. This process maintains the noise in the image, and the pixel value of the input image is mathematically defined in Equation (5.1).

$$\varkappa_{(\mathfrak{o},\mathfrak{l})} = \mathcal{N}_{\mathfrak{ab}}(\mathfrak{o},\mathfrak{l}) \sum_{\mathfrak{x}=-3\sigma}^{3\sigma} \delta_{\mathfrak{ab}}(\mathfrak{o},\mathfrak{l},\mathfrak{u},\mathfrak{v})\mathcal{I}(\mathfrak{o}+\mathfrak{u},\mathfrak{l}+\mathfrak{x}) \qquad (5.1)$$

Let $\mathcal{N}_{\mathfrak{ab}}$ represent a normalization bilateral filter, and $\delta_{\mathfrak{cb}}$ denote the dimensions corresponding to each pixel in the filter space. The bilateral filter is employed to achieve the smoothing function $\mathfrak{f}_{\mathfrak{z}}(\mathfrak{u}, \mathfrak{v})$ for estimating the Gaussian size. Introducing a Gaussian weight factor through a visual inclusion factor enables the flexible selection of pixel values in the scaled overall form $(\mathfrak{o}, \mathfrak{l}, \mathfrak{u}, \mathfrak{v})$.

This, in turn, yields the expression for the bilateral size estimation function as given in Equation (5.2).

$$\delta_{\mathfrak{ab}}(\mathfrak{o}, \mathfrak{l}, \mathfrak{u}, \mathfrak{v}) = \mathfrak{f}_{\mathfrak{z}}(\mathfrak{u}, \mathfrak{v}) \tag{5.2}$$

$\mathcal{N}_{\mathfrak{ab}}$ is a normalization factor that is determined by taking the reverse of the total dimensions. It translates the sum of the bilaterally dimensioned forces into an average function. $\mathcal{N}_{\mathfrak{ab}}$ is demonstrated in Equation (5.3).

$$\mathcal{N}_{\mathfrak{ab}}(\mathfrak{o}, \mathfrak{l}) = \frac{1}{\sum_{\mathfrak{x}=-3\sigma}^{3\sigma} \sum_{\mathfrak{x}=-3\sigma}^{3\sigma} \delta_{\mathfrak{ab}}(\mathfrak{o}, \mathfrak{l}, \mathfrak{u}, \mathfrak{v})} \tag{5.3}$$

The bilateral filter serves as a nonlinear filter for reducing noise. Filtering and pixel-level considerations impact the outcomes. Equation 5.4 is employed to calculate the range of filter pixels, taking into account their distance from the point of origin.

$$\mathfrak{g}(\mathfrak{o} - \mathfrak{l}) = \frac{1}{2}\mathcal{Q}^{-(\mathfrak{o}-\mathfrak{l})^{\wedge}2(\frac{1}{2\tau})} \tag{5.4}$$

From the above equation, $\mathfrak{o} - \mathfrak{l}$ be the pixel value of the image position, and 2τ represents the spatial function. The range filter assigns weights to pixels based on the degree of sound variation, as indicated in Equation (5.5).

$$\mathfrak{T}(\mathfrak{h}(\mathfrak{k}) - \mathfrak{h}(\mathfrak{l})) = \frac{1}{2}\mathcal{Q}^{\frac{(\mathfrak{h}(\mathfrak{k})-\mathfrak{h}(\mathfrak{l}))(\mathfrak{h}(\mathfrak{k})-\mathfrak{h}(\mathfrak{l}))}{2\tau_{\varsigma}}} \tag{5.5}$$

From the equation, $\mathfrak{h}(.)$ represent the pixel function of the input image, and $2\tau_{\varsigma}$ denotes the filtering set of the image. The images devoid of noise are inputted into the HE filters to improve the quality of the initial images.

5.2.2 Histogram Equalization

Pre-processing is conducted on gathered images to safeguard crucial information in the LV. The output pixel value is generated by linearly combining the pixel factors of the input through linear filters. The acquired images undergo initial processing utilizing an ABIHE filter, aiming to enhance image quality while preserving vital information. ABIHE filters prove to be efficient in refining and improving snapshots, particularly when confronted with images exhibiting poor contrast, noise, and variable local characteristics. This preprocessing involves tasks such as noise reduction, picture denoising, and scaling. The objective of these operations is to enhance the overall quality of the final product. Such techniques are crafted to elevate the quality of the end result and ready the image for subsequent processing.

The ABIHE filtering operates on pixels located in the immediate vicinity of each pixel in the image. The ABIHE filter assigns weights to assess the impact of neighboring pixels on the processed pixel, and these weights are employed to calculate the weighted average of pixel values within the neighborhood. Image processing employs histogram equalization as a method to enhance contrast by redistributing pixel intensities. Although beneficial in certain image analysis scenarios, it's not universally required. The Histogram Equalization Module individually equalized each sub-histogram using specific equations. The ultimate output image results from the summation of all generated sub-images.

$$\mathfrak{D} = \begin{cases} \mathfrak{e} * \ell \; 0 \leq \mathfrak{w} < z \\ \mathfrak{f} * (\mathfrak{w} - \mathfrak{t}) + \mathfrak{d} \; x \leq \mathfrak{w} < z \\ \mathfrak{g} * (\mathfrak{w} - \mathfrak{o}) + \mathfrak{b} \; y \leq \mathfrak{w} < e - 1 \end{cases} \tag{5.6}$$

As mentioned in Equation 5.6 the pixel intensity values $\mathfrak{e}$, $\mathfrak{f}$, $\mathfrak{g}$ and$\mathfrak{d}$ to encompass the entire dynamic range of the images. Let $\mathfrak{b}$ represents the input image, and the calculated intensity range defines the input range. Presently, the improved images are inputted into the Region of Interest (RoI) to extract pertinent features from the image for segmentation.

5.3 REGION OF INTEREST (ROI) EXTRACTION

ARoI extraction technique is employed to detect left ventricle segmentation. The RoI extraction method is utilized to segment the LV, as progressive heart diseases resulting from cardiovascular injury. Subsequently, features are extracted from raw pixel intensities through a three different deep learning approaches. Following feature extraction, test images are segments. The ROI refers to an area within an image that proves useful for specific objectives. ROI extraction is crucial for reducing computation time and costs while enhancing accuracy. In the context of CMRI imaging, the region surrounding and encompassing the disc is identified as the ROI of a LV image. R-CNN represented a significant leap forward in the realm of object detection, playing a crucial role in advancing the field. It stood out as one of the pioneering models that demonstrated the capabilities of deep learning in handling object detection tasks. The R-CNN architecture consists of three key stages: the backbone, Region Proposal Network (RPN), and RoI. Convolutional techniques are employed on input images to extract the relevant conceptual information. During the initial stage, the RPN generates suggestions that may encompass foreground elements, while in the subsequent step, the RoI network refines the proposal results and performs segmentation. Both the RPN and RoI networks leverage the convolutional feature maps as their foundational components for these processes.

Left ventricle segmentation in noise-free CMRI involves a three-stage RoI-based extraction process to achieve accurate identification. Initially, a bounding box is created around the LV in the first stage, followed by the application of a Canny edge detector to mask the contour edges of the LV. Subsequently, contour points are identified through mask thresholding. These extracted features serve as input for three distinct deep learning structures, collectively referred to as Tri-ConvNets, aimed at efficiently segmenting the LV. Following the rectification of the image region by the RoI network stage, RoI Align was utilized to obtain improved feature maps. These enhanced feature maps were then inputted into the mask layer for image segmentation. Equations (5.7) and 5.8 represent the RPN layer and the RoI layer, respectively, and these layers contribute significantly to the complexity in the loss variation within the RCNN. The loss value of the RPN is formed by adding the regression and region-segmented losses.

$$RPN_{ls} = \sum_{i=1} \mathcal{Q}_l(\mathcal{N}_i) + \mathcal{P}_l(\mathcal{N}_i) \tag{5.7}$$

The RoI layer's loss rate has been split up into three components: segmentation loss L_{los}, regression loss $\mathcal{P}_l$, and classification loss $\mathcal{E}_l$. A feature map with different scales will be combined bilinearly by the RoI method to create a uniform scale. The RoI layer's overall loss is calculated as follows,

$$ROI_l = \mathcal{Q}_l + \mathcal{P}_l + \mathcal{E}_l \tag{5.8}$$

Bilinear interpolation will be performed by the RoI procedure to transform a feature map with different scales into one scale. The segmentation level is also in charge of segmenting the region of interest. To achieve extremely high accuracy in ROI recognition on the datasets implemented.

5.3.1 Extraction of Bounding Boxes through Region of Interest (ROI)

In medical image analysis, the extraction of the LV is a vital phase for various applications, such as quantitative analysis and clinical diagnosis. The bounding box pattern in left ventricle extraction refers to the use of bounding boxes to define and extract the ROI comprising the left ventricle from medical images. Before extracting the left ventricle, preprocessing steps applied to enhance the quality of the images. Once the approximate location is identified, a bounding box is defined around the region of interest in left ventricle. A bounding box is a rectangular area that encloses the target object. The coordinates of the bounding box $(\mathfrak{o}, \mathfrak{p}, \mathfrak{q}, \mathfrak{r})$ are determined based on the detected location. The pixels within the bounding box are then extracted to form the ROI. This step essentially isolates the left ventricle from the rest of the image. Once the bounding box is generated for the left ventricle, the contour edge and contour points are estimates using the RoI extraction approach. The bounding box pattern simplifies the process of left ventricle extraction by providing a well-defined region for analysis.

5.3.2 Extraction of Contour Edges through ROI

The contour of an object in an image refers to the outline or boundary that separates it from the background. The contour edge pattern involves the analysis of the patterns formed by the edges of the object. The Canny edge detector is an algorithm used in image processing for edge detection. It is employed in the extraction of the LV from cardiac images. Sensitivity to image noise is a notable characteristic of edge detection. To address this, the image is concatenated with a Gaussian kernel $(3x3,\ 5x5,\ 7x7)$, the core size depending on the expected blur trace. Essentially, the blur is less visible for the smallest kernel. Converting the input image to grayscale by adjusting contrast and brightness is necessary to blur the image and eliminate noise. To make edge location and detection effective, a filter is utilized to remove noise in the main image. The Gaussian filter is commonly used for this purpose and is expressed as follows.

$$\mathfrak{g}(\mathfrak{X}, \mathfrak{y}) = \frac{1}{2\pi\tau^2} \exp\left(-\frac{\mathfrak{x}^2+\mathfrak{y}^2}{2\tau^2}\right) \tag{5.9}$$

From the above Equation (5.9), $\mathfrak{y}$ specifies the axial position of source, $\mathfrak{y}$ represents the axial location, and τ is the Gaussian's variance.

The Canny edge detection operator identifies both the edges and their directional intensities. It achieves this by computing the image gradient, pinpointing edge pixels with significant changes in gray area values. The points in the direction of the most intensity variation is represented by the gradient, which is normalized as a unit vector. Initially, the vertical and horizontal components of the gradient are determined, followed by the computation of the gradient's magnitude and orientation. The calculations for gradient magnitude (Υ) and gradient angle ($\mathfrak{z}$) proceed as follows:

$$\Upsilon = \sqrt{\mathfrak{z}_{\mathfrak{x}}^2 + \mathfrak{z}_{\mathfrak{y}}^2} \quad (5.10)$$

$$\mathfrak{z} = \tan\left(\frac{\mathfrak{z}_{\mathfrak{y}}}{\mathfrak{z}_{\mathfrak{x}}}\right) \quad (5.11)$$

From the above Equation (5.10) and (5.11), $\mathfrak{z}_{\mathfrak{y}}$, and $\mathfrak{z}_{\mathfrak{x}}$ demonstrate the angles of the image. Finally, the Canny edge detection algorithm utilizes to detect contour edge of LV.

5.3.3 Extraction of Contour Points through ROI

The contour points are detected after detecting contour edges. Contour points play a crucial role in extracting and delineating the outer border region of the left ventricle images. In the context of left ventricle image extraction, these points represent the boundary of the left ventricle within the image. These points are usually obtained through image processing techniques that identify edges and contours within the image. Various image processing techniques extract contour points and outline the left ventricle. These identified points form a closed curve representing the ventricle's boundary. Crucial for image extraction, contour points define the ventricle's boundaries, essential for quantitative analysis and diagnoses in cardiac imaging. Boundary boxes, contour edges, and contour points are extracted using a RoI extraction model, and these features are fed into three deep learning networks for LV segmentation.

5.4 TRI-CONVOLUTIONAL NETWORKS (CONVNETS)

The extracted features are fed input to the three different deep learning structures for segmenting the LV in the efficient way. Generalized characteristics are extracted from cardiac MR images using a Transformer, and coarse-grained semantics and fine-grained features are extracted from complete scales using Tri-ConvNets. It resolves the issue of poor segmentation accuracy brought on by hazy LV edge data. The contour edges are processed in the standard ConvNet, the contour points are processed using Fully ConvNet and finally the noise free images are converted into patches to perform the pixel-wise operation in ConvNets. These three ConvNets performs different operations to give different segmentation results and these results concatenate to calculate the circular similarity for attaining better LV segmentation output.

5.4.1 Standard Convnets

Contour edge segmentation using Standard CNN involves the extraction of boundaries and edges from an input image. A standard CNN architecture as given in the Figure 5.3is used for this task typically consists of several layers. The convolutional layers play a vital part in capturing local patterns and features. Forecasts using feature learning are aided by fully linked layers at the final stage of the structure.

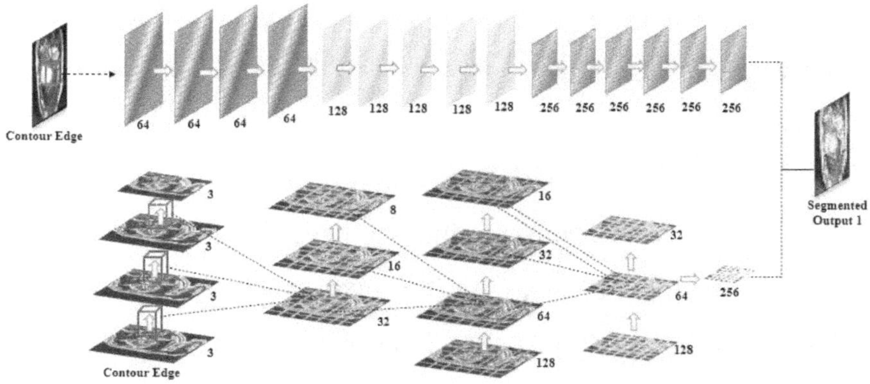

Figure 5.3 Architecture of standard convolutional network

Let αspecifies the input image and γ denotes the output image. The convolutional operation is specified in equation (). The dimension of the matrix is denoted as $\mathfrak{z}$, and the bias of the image denoted as $\mathfrak{p}$. The below represented operation is a non-linear activation factor. The ReLU function is denotes as $Q' = \mathfrak{f}(Q)$. The output of the minimum value within a local region is expressed as $\mathbb{A} = max_{pl}(Q')$ and is given in the Equation (5.12).

$$Q = \alpha \times \mathfrak{z} + \mathfrak{p} \tag{5.12}$$

For contour edge segmentation, the final layers of the network are designed to produce a binary segmentation map indicating the presence or absence of edges. The output can be obtained using a sigmoid activation function.

$$\mathfrak{z}' = \pi(\mathfrak{g}' \cdot \mathfrak{a} + \mathfrak{p}') \tag{5.13}$$

In the above Equation (5.13) the expression, $\mathfrak{g}'$ and $\mathfrak{p}'$ specifies the weight and bias of the final layer and π be the sigmoid function. The loss function used during training is frequently the binary cross-entropy loss that trials the dissimilarity among the forecast output and the ground truth segmentation map. The standard CNN for contour edge segmentation involves the application of convolutional and pooling operations to extract hierarchical features, followed by fully connected layers to make final predictions. The system is qualified to minimize the binary cross-entropy loss, ensuring accurate edge segmentation in the outcomes.

5.4.2 Fully CNN

A FCN for contour points segmentation is a deep learning architecture designed as in Figure 5.4 to address pixel-wise segmentation tasks. FCNs are particularly well-suited for such tasks as they preserve spatial information through the use of convolutional layers. The key innovation of FCNs is their ability to handle input images of arbitrary sizes and produce output segmentation maps with the same spatial dimensions as the input.

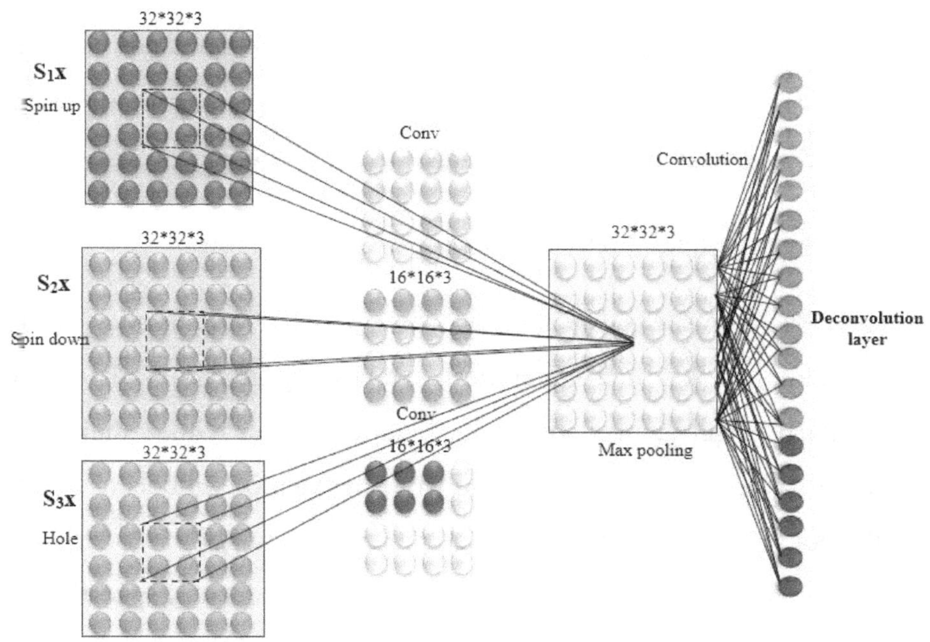

Figure 5.4 Architecture of fully ConvNet

Let, Υ represent the input image, and ξ denote the ground truth map. The encoder consists of convolutional layers $\mathcal{F}_{\epsilon}$ that encode spatial information, and the decoder consists of transposed convolutional layers $\mathcal{F}_{\mathfrak{d}}$ that up sample the feature maps. The final segmentation map is generated by applying a SoftMax activation function, denoted as $\mathcal{F}_{\mathcal{S}}$, to obtain class probabilities for each pixel and is expressed in Equation (5.14).

$$\xi = \mathcal{F}_{\mathcal{S}}(\mathcal{F}_{\mathfrak{d}}(\mathcal{F}_{\epsilon}(\Upsilon))) \tag{5.14}$$

For training purposes, a cross-entropy loss rate generally utilized, which measures the dissimilarity between the predicted segmentation map ξ and the ground truth ξ.

$$\mathcal{Q}(\hat{\xi}\xi) = -\frac{1}{e}\sum_{p=1}^{\varrho}\sum_{q=1}^{\varrho}\xi_{p,q}\log(\hat{\xi}_{p,q}) \tag{5.15}$$

From the above evaluation 5.15, e be the sum of pixels, ϱ is the sum of classes, $\xi_{p,q}$ and $\hat{\xi}_{p,q}$ represent the ground truth and predicted probabilities for pixel p and q, respectively. After segmenting the images with fully convolutional networks, contest patches are generated using pixelwise CNNs.

5.4.3 Pixelwise CNN

Pixelwise CNN are a class of DL models designed for image segmentation tasks, such as contest patches segmentation. These networks are particularly effective in capturing local dependencies within images, making them suitable for tasks where the relationship between neighboring pixels is crucial. In the context of contest patches segmentation, the goal is to classify each pixel into distinct classes, indicating the presence or absence of specific features in the image. The architecture of a Pixel CNN as shown in Figure 5.5 involves stacking multiple convolutional layers to progressively learn hierarchical representations of the input image.

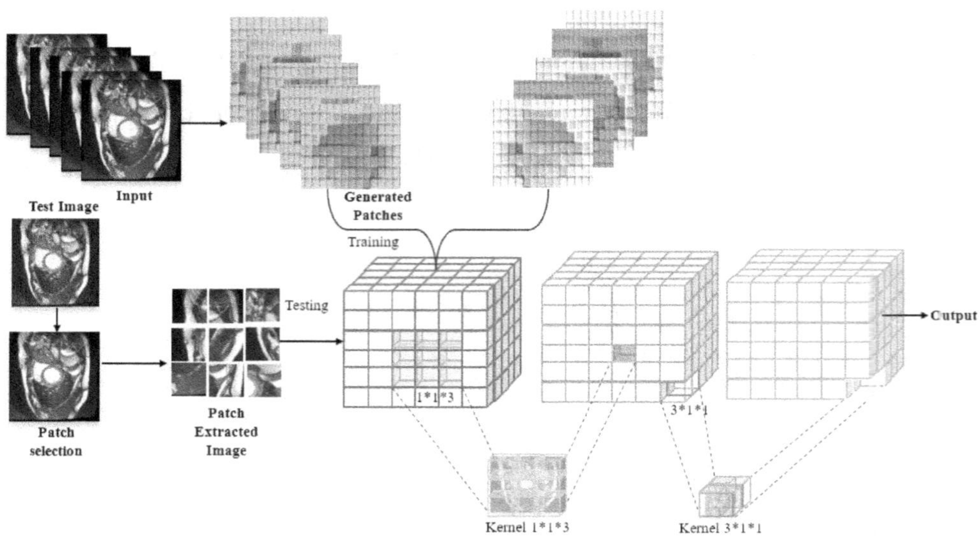

Figure 5.5 Architecture of pixel wise convolutional network

Each convolutional layer employs a receptive field that captures information from a local neighbourhood of pixels. The output of a pixelwise convolutional layer can be expressed in Equation (5.16).

$$\mathfrak{d}'_{\mathfrak{p},\mathfrak{q}} = \xi(\omega \times \mathfrak{d}_{\mathfrak{p},\mathfrak{q}} + \varsigma) \tag{5.16}$$

From the above equation, $\mathfrak{d}'_{\mathfrak{p},\mathfrak{q}}$ denotes the input pattern at pixel location $(\mathfrak{p},\ \mathfrak{q})$, ω and ς denotes the convolutional filter parameters and bias factor. For contest patches segmentation, the final layer of the Pixel CNN is often a SoftMax layer, which outputs probability distributions over different classes for each pixel. The segmentation task for pixelwise classification problem, and the cross-entropy loss is commonly utilized to evaluate the dissimilarity among predicted and ground truth probability distributions and is expressed in Equation (5.17).

$$\delta = -\frac{1}{\Gamma}\phi_{\mathfrak{p},\mathfrak{q}}\phi_o \upsilon_{\mathfrak{p},\mathfrak{q},o} \log(\hat{\mathfrak{y}}_{\mathfrak{p},\mathfrak{q},o}) \tag{5.17}$$

From the above equation, Γ is the sumr of pixels, $\upsilon_{\mathfrak{p},\mathfrak{q},o}$ is the ground truth probability of class o at pixel $(\mathfrak{p},\ \mathfrak{q})$, and $\hat{\mathfrak{y}}_{\mathfrak{p},\mathfrak{q},o}$ is the predicted probability of class at $o$atthe same pixel.The network is trained to minimize this loss function, optimizing its parameters to accurately segment contest patches based on the input images.

5.5 SUMMARY

In this chapter it summarizes about the architecture of Tri-Convolutional Networks and its algorithm is explained. The concepts of CNN and Pixelwise CNN is elaborated in a detailed manner. The Region of Interest Extraction of the images used in Tri-Convolutional Networks is explained in this chapter.

CHAPTER 6

RESULTS AND DISCUSSION

The experiment findings are analyzed in this section, and the efficiency of both the proposed model is evaluated and compared with existing state-of-the-art techniques.

6.1 DATASET DESCRIPTION FOR ES-FCN

The datasets used in our experiments are summarized in Table 6.1, four different datasets of short-axis MR images were used here.

Table 6.1 Datasets summary

Dataset	No of Patients	No of Images	No of Labelled Images	No of Slices/Patient	No of Frames/slice
ACDC	100	25351	1902	6 – 18	2
LVSC	100	29086	2522	8 – 24	2
Sunnybrook	45	51545	420	5 – 6	1 – 2
York	33	7980	7980	8 – 15	20

- The Automated Cardiac Diagnosis Challenge (ACDC) dataset contains short-axis MR images from 100 subjects. The images are in NIfTI (Neuro Imaging Informatics Technology Initiative) format.

The ground-truth segmentation boundaries are manually collected from the experts for both LV and RV cavities. The spatial resolution of the images varies from 1.36 to 1.68 mm^2/pixel and 28 to 40 images cover the cardiac cycle partially or completely.

- The Sunnybrook (SB) dataset contains cine-MRI images from 45 patients, who were diagnosed from different pathologies group. The images are in the DICOM (Digital Imaging and Communications in Medicine) format, where the images are stored along with metadata.
- The LV Segmentation Challenge (LVSC) dataset comprises short-axis MRI images obtained from 100 patients. The dataset was prepared in the year 2011 for a study called DETERMINES (Defibrillators to Reduce Risk by Magnetic Resonance Imaging Evaluation). Image resolutions range from 138×192 to 512×512.
- The York dataset contains 7980 MRI images, acquired from 33 patients, the dataset has been prepared by the Department of Diagnostic Imaging of the Hospital for Sick Children in Toronto, Canada. All the patients were under the age of 18. Each slice of the image follows a resolution of 256×256.

6.2 EVALUATION METRICS FOR ES-FCN

The segmentation accuracy is estimated against the ground truth and the efficiency was evaluated with six accuracy measures: Sensitivity (p), specificity (q), Positive Predictive Value (PPV), Negative Predictive Value (NPV), Jaccard index ($\mathcal{J}$) and Dice index ($\mathcal{D}$). The measures are computed in Equation (6.1) – (6.10).

$$p = \frac{T_1}{N_1} \tag{6.1}$$

$$q = \frac{T_0}{N_0} \tag{6.2}$$

$$PPV = \frac{T_1}{T_1+F_1} \tag{6.3}$$

$$NPV = \frac{T_0}{T_0+F_0} \tag{6.4}$$

where, T_0 and T_1 denote the number of correctly segmented pixels as Left Ventricle regions and non-LV regions, while F_1 and F_0 represent the number of miscategorized pixels segmented as LV and non-LV regions, respectively. The number of LV and non-LV pixels in the ground truth region a is represented with N_1 and N_0 respectively.

The Dice and Jaccard indices are the popular measures used for determining segmentation results. The Dice index is computed as in Equation (6.5).

$$\mathcal{D}(D_1, T_1) = \frac{2|D_1 \cap T_1|}{|D_1|+|T_1|} \tag{6.5}$$

and the Jaccard index is given as in Equation (6.6),

$$\mathcal{J}(D_1, T_1) = \frac{|D_1 \cap T_1|}{|D_1 \cup T_1|} \tag{6.6}$$

where, D_1 and $T_1$1 reflect the number of pixels from the predictable segmentation regions and the ground-truth regions respectively. The indices with a value close to 1 indicate better segmentation performance.

6.3 RESULT ANALYSIS OF PRE-PROCESSING STAGE

A discussion of the subsequent results for each phase of the proposed ES-FCN model will be presented in this section. A binary convolution layer-based ES-FCN model is created to reduce the complexity of traditional FCN models. Pre-processing steps including edge detection, modification, and morphological procedures are used to create binary images. Results from the data augmentation step are shown in examples in Figure 6.1. The input images are transformed to support the first FCN layer in the ES-FCN model.

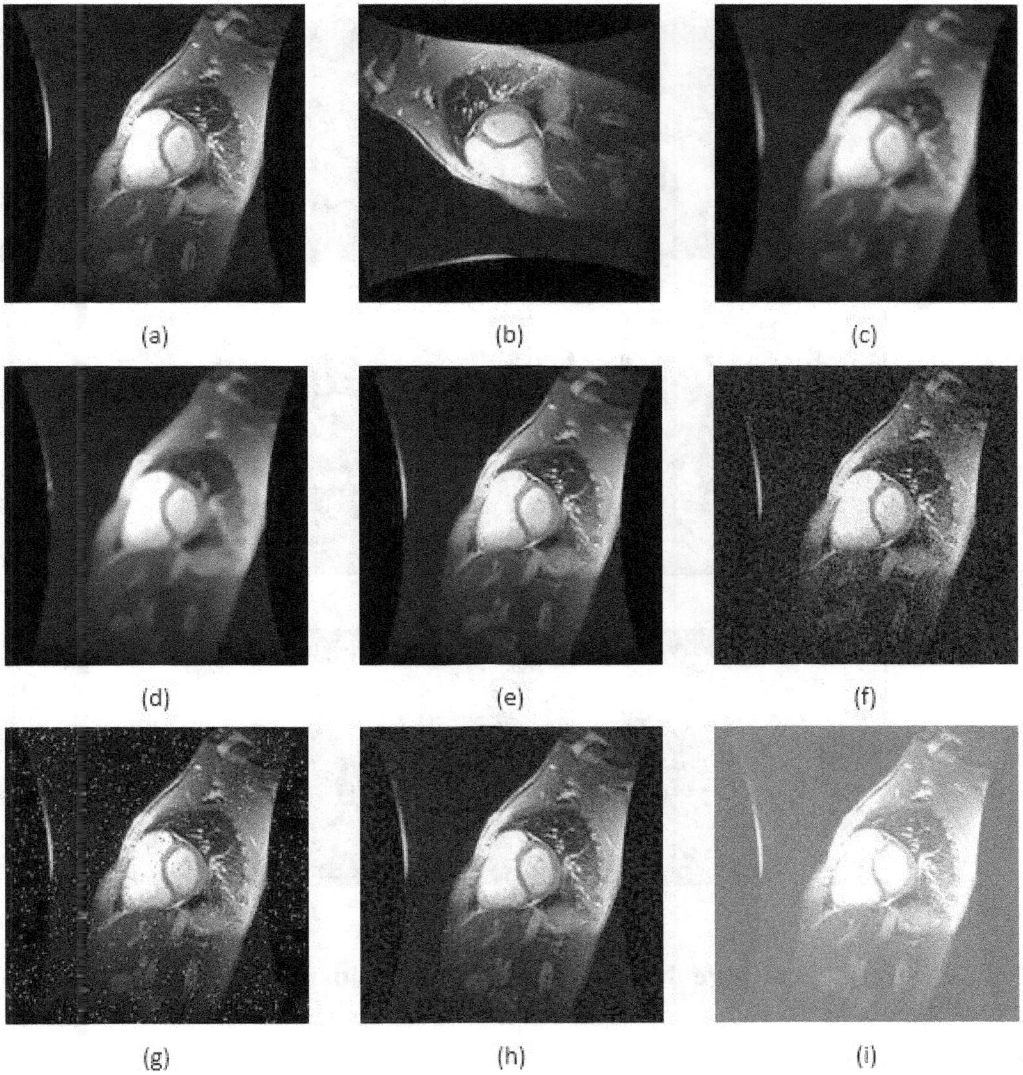

Figure 6.1 The input images to ES-FCN after image manipulators (a) MR-image, (b) Image rotated with 45° angle, (c) Blurred image after Box filter, (d) Smoothened image with median filter, (e) Smoothened image with wiener filter; Images degraded with (f) Gaussian noise, (g) Salt and pepper noise, (h) Gamma noise and (i) Uniform noise

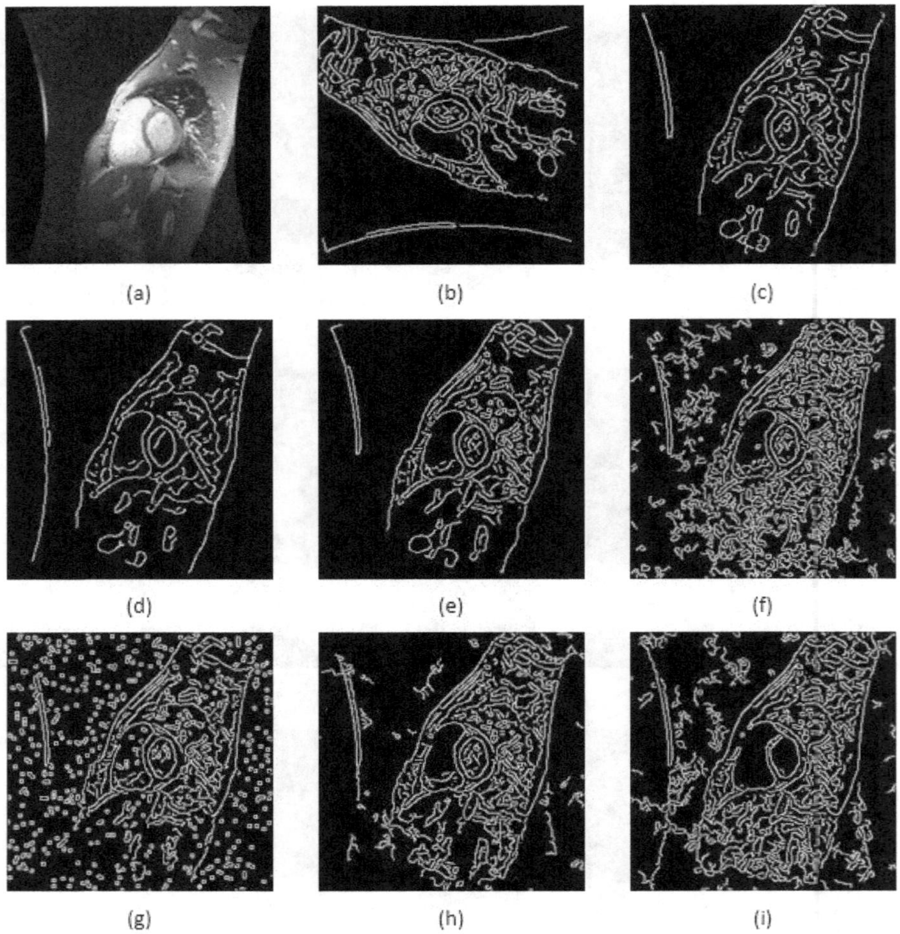

Figure 6.2 Edge detection in MR images using a canny edge detector

The results of the fuzzy-based canny edge detection system are displayed in Figure 6.2 by manipulating the edges in the MR images. This methodology, in contrast to conventional edge detection, can deal with the ambiguity of LV structures. The outcomes of the morphological open process procedure, which fills the regions of the ventricle and atrium, are shown in Figure 6.3.

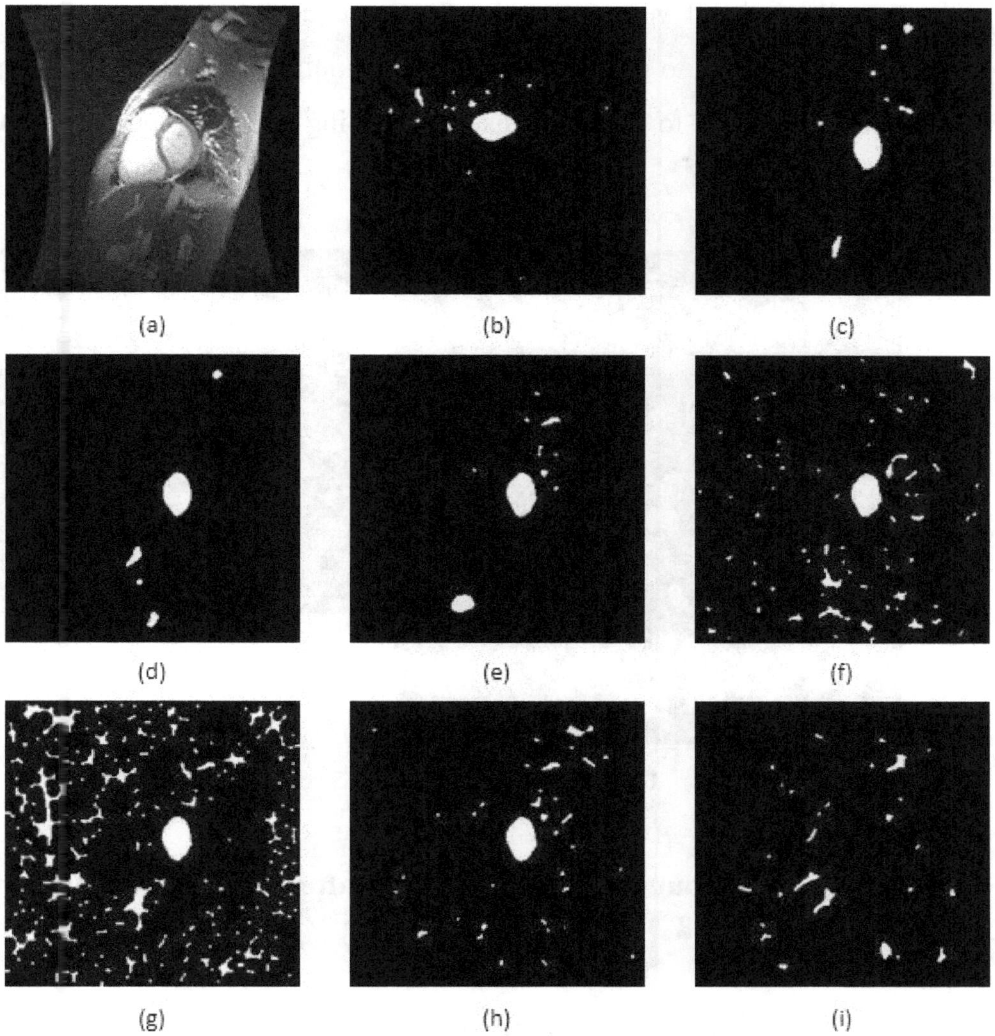

Figure 6.3 Experimental results after a morphological open operation

6.4 RESULT ANALYSIS OF THE ES-FCN MODEL

This work presents a novel ES-FCN-based segmentation model that combines bicubic interpolation-based decoding, max pooling, and binary convolution. On the input images, the encoder performs numerous different-level feature representations. The encoder is built in the ES-FCN model, and its structure allows the model to understand context and temporal dependencies of the sequences. In the hidden state, the encoder output is the state of the network timestep. In the machine learning model, the role of the decoder will be to convert the two-dimensional vector into the output sequence. In the network, bicubic interpolation is used to generate the network input as a preprocessing step, which provides a solution to the inconsistent mapping problem in the ES-FCN and encoding algorithms.

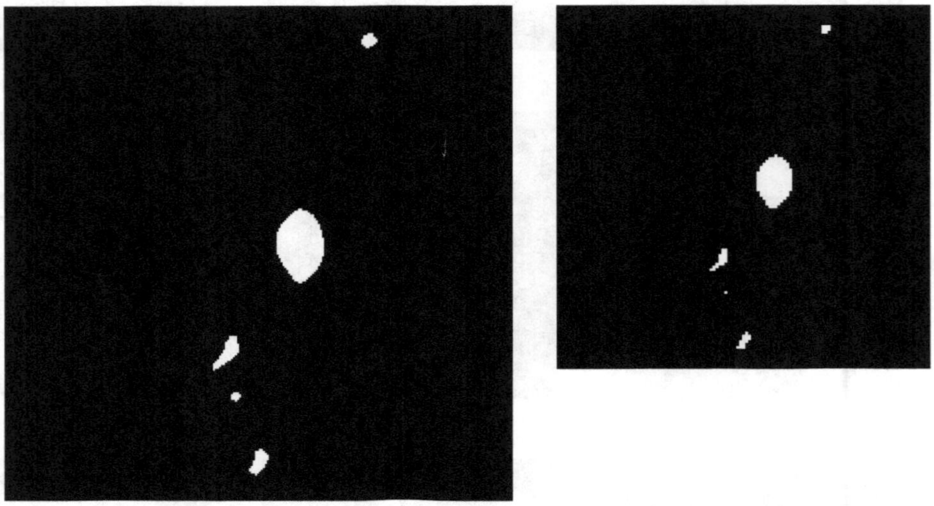

(a) (b)

Figure 6.4 (a) Input image and (b) Image after one-step convolution and pooling

(a) (b)

Figure 6.5 Convoluted and pooled images after (a) Second and (b) Third steps

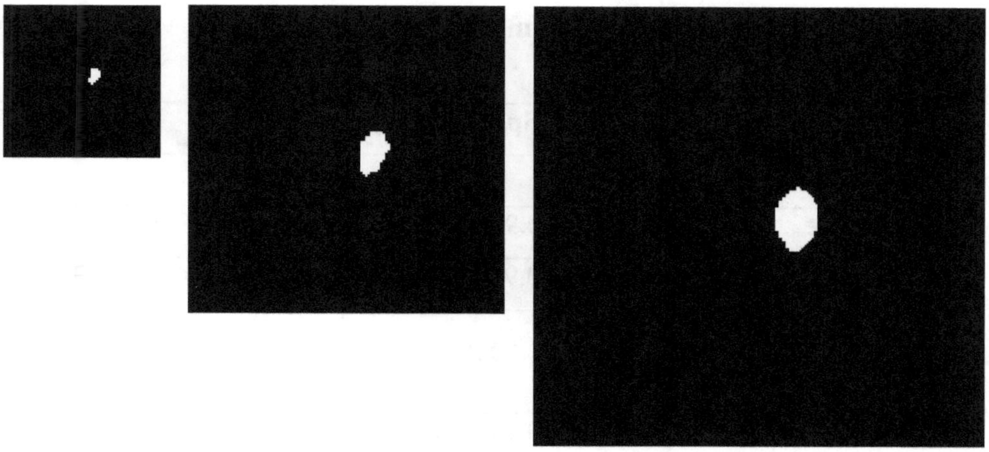

(a) (b) (c)

Figure 6.6 Deconvolution outputs (a) First, (b) Second and (c) Third level

Figures 6.4 and 6.5 display the encoder layer results for the suggested ES-FCN models. In Figure 6.4 the images from the median filtered are taken as input to the proposed network. These inputs are processed in each convolutional and pooling layer that is represented in 6.4 (b) first convoluted and pooled output 6.5 (a) convoluted and pooled output, and 6.5 (c) convoluted and pooled output The convoluted and pooled output of the third layer is deconvoluted to its original size as shown in Figure 6.6. Figure 6.6 simultaneously displays the decoder layer findings; bicubic interpolation overlaps the multiscale features.

6.5 EVALUATION OF ACDC DATASET

Tables 6.2 & 6.3 quantify the segmentation performance measures estimated of the short-axis MR images from the ACDC dataset. The average and variance values of the performance measures are depicted. The experimental result indicates that the proposed ES-FCN segmentation approach performs better than the other models with higher accuracy. The r PPV of 0.9595 and higher Jaccard index of 0.9484 indicate the effective and sensitive LV segmentation from the ES-FCN method.

Table 6.2 Comparison of segmentation accuracy with MR images from the ACDC dataset

Algorithm	Sensitivity (p)	Specificity (q)	PPV	NPV
Level set	0.943	0.9216	-	-
MA-Shape	0.9332±0.0305	0.9919±0.0007	0.9353±0.0376	0.9962±0.0376
DLDP	0.9179±0.0344	0.9943±0.0052	0.9272±0.0423	0.9957±0.0423
FCN (traditional)	0.7577±0.0165	0.7949±0.0015	0.7595±0.0237	0.7976±0.0237
ES-FCN	0.9577±0.0165	0.9949±0.0015	0.9595±0.0237	0.9976±0.0237

Table 6.3 Segmentation validation on ACDC dataset with dice and jaccard indices

Algorithm	Dice Index ($\mathcal{D}$)	Jaccard Index ($\mathcal{J}$)
Level set	0.885	0.965
2D U-Net	94.73	-
MA-Shape	0.9344±0.0328	0.9468±0.0328
DLDP	0.9334±0.0376	0.9464±0.0376
MMNet	0.951	0.901
FCN (conventional)	0.9277±0.0470	0.9454±0.0470
ES-FCN	0.9319±0.0188	0.9484±0.0188

In addition to the performance measures, linear regression plots are used to establish the degree of comparison of automated and ground-truth segmentation. To improve the readability of the illustration, the plot is generated only with the first 500 samples from the dataset. Figure 6.7, depicts the segmentation comparison between the number of pixels received from the automated segmentation and the physical segmentation. The fitness curve specifies the superior performance of the proposed ES-FCN model.

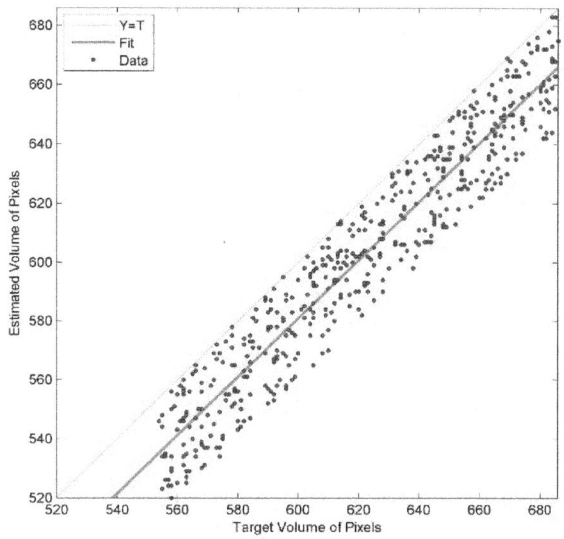

Figure 6.7 Correlation plot showing automatic segmentation volumes with physically attained volumes using MR images from the ACDC dataset

6.6 EVALUATION OF LVSC DATASET

Tables 6.4 & 6.5 presented the segmentation accuracy from the LVSC dataset. The experimental result indicates that the proposed ES-FCN segmentation model is superior to the other methods with greater accuracy. The better PPV of 0.9592 and greater Jaccard index of 0.9476 indicate the significant LV segmentation performance from the ES-FCN method. Figure 6.8 illustrates the higher similarity between automated and manual segmentation.

Table 6.4 Comparison of segmentation accuracy with MR images from the LVSC dataset

Algorithm	Sensitivity (p)	Specificity (q)	PPV	NPV
AGU-Net	0.87	0.92	0.87	0.93
MA-Shape	0.9310±0.0285	0.9952±0.0011	0.9389±0.0355	0.9965±0.0355

Table 6.4 Continued

DLDP	0.9243±0.0324	0.9941±0.0066	0.9316±0.039 6	0.9961±0.0396
FCN	0.83	0.96	0.86	0.95
FCN (conventional)	0.72	0.713	0.6934	0.738
ES-FCN	0.9538±0.0162	0.9956±0.0059	0.9592±0.0237	0.9977±0.0237

Table 6.5 Segmentation validation on LVSC dataset with dice and jaccard indices

Algorithm	Dice Index ($\mathcal{D}$)	Jaccard Index ($\mathcal{J}$)
AGU-Net	0.937	0.75
MA-Shape	0.9343±0.0355	0.9466±0.0355
DLDP	0.9327±0.0396	0.9462±0.0396
FCN	0.94	0.74
FCN (conventional)	0.74	0.725
ES-FCN (proposed)	0.9314±0.0237	0.9476±0.0237

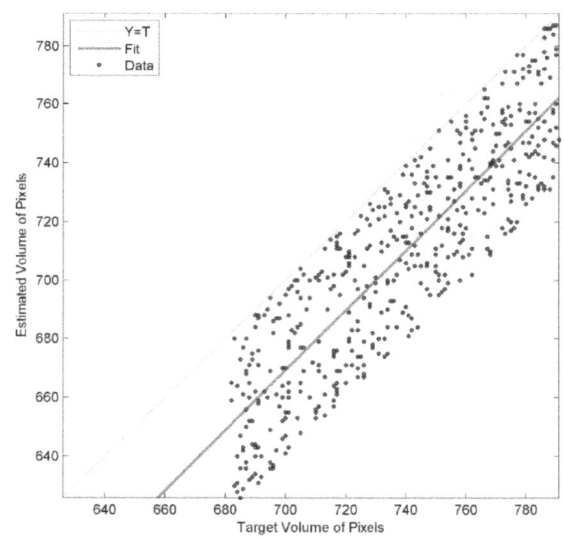

Figure 6.8 Correlation plot showing automatic segmentation volumes with physically attained volumes using MR images from the LVSV dataset

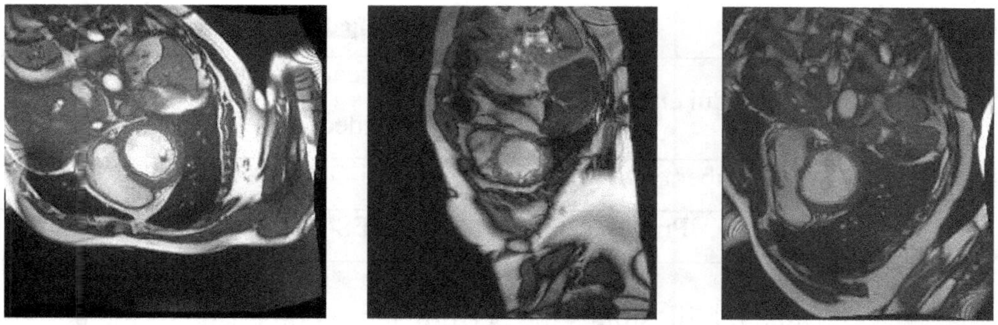

(a)

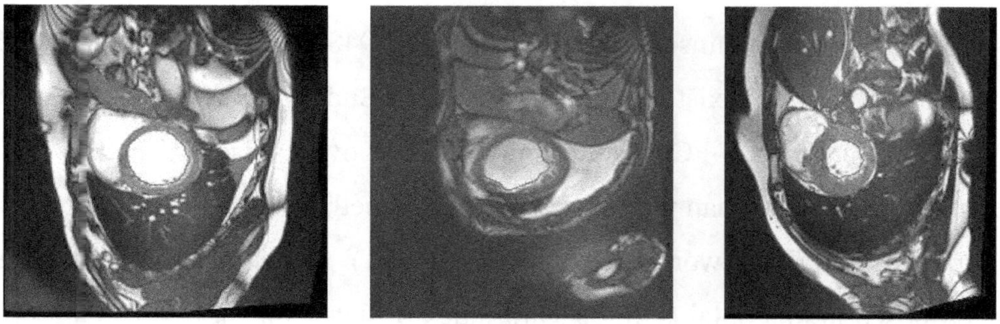

(b)

Figure 6.9 Sample segmentation results of (a) ACDC dataset and (b) LVSV dataset

The final segmentation outcomes of the ES-FCN model on the ACDC and LVSV datasets are shown in Figure 6.9. The aforementioned conclusions show that the model we've proposed is an effective technique for helping cardiologists screen for illnesses. This approach is ideal for real-time applications due to its low level of complexity and computing.

Table 6.6 Comparison analyses between the proposed and the existing models based on ACDC and LVSV datasets

Datasets	Authors	Model	Jaccard Index
ACDC	Budai *et al.* (2020)	Regression-based method	0.84
	Guo*et al.* (2021)	Pre-trained neural network	0.81
	Proposed	ES-FCN	0.94

Table 6.6 Continued

LVSV	Cui *et al.* (2021)	Multiscale attention guided U-Net	0.75
	Saber *et al.* (2020)	Multi-Gate U-Net	0.82
	Proposed	ES-FCN	0.94

Table 6.6 illustrates the performance comparison of the proposed ES-FCN model with the existing methods in terms of Jaccard indices. The ES-FCN model attains the Jaccard index of 0.948 for the ACDC dataset which is greater than the existing models such as Budai *et al.* (2020) and Guo *et al.* (2021). The ES-FCN model improves the overall Jaccard indices by 10.6% and 13.8% better than the Regression-based method Budai *et al.* (2020) and Pre-trained neural network Guo *et al.* (2021) for the ACDC dataset. The proposed model obtains the Jaccard index of 0.947 for the LVSV dataset, which is greater than the existing models such as Cui *et al.* (2021) and Saber *et al.* (2020). The ES-FCN model improves the overall Jaccard indices by 20.2% and 12.7% better than the Multiscale attention-guided U-Net Cui *et al.* (2021) and Multi-Gate U-Net Saber *et al.* (2020) for the LVSV dataset. From the table, the proposed ES-FCN model yields a higher Jaccard index than the existing methods for both datasets indicating that the proposed model is superior to the prior models.

6.7 DESCRIPTION OF DATASET FOR TRI-CONVNETS

The Sunnybrook and York dataset is implemented for the segmentation of LV. The Sunnybrook Cardiac Data is a dataset used in medical imaging research, particularly in the field of cardiac image analysis. It is widely employed for tasks such as LV segmentation and motion estimation in cardiac MRI scans. Table 6.11& 6.12 listed the segmentation accuracy from Sunnybrook dataset.

The Table 6.8 & 6.9 listed the segmentation accuracy from York dataset. The study's experimental setup utilized Spyder, an Anaconda navigator running on a Windows 10 operating system. The PC employed an Intel i5 core processor with a speed of 2.10 GHz and a 16GB RAM system.

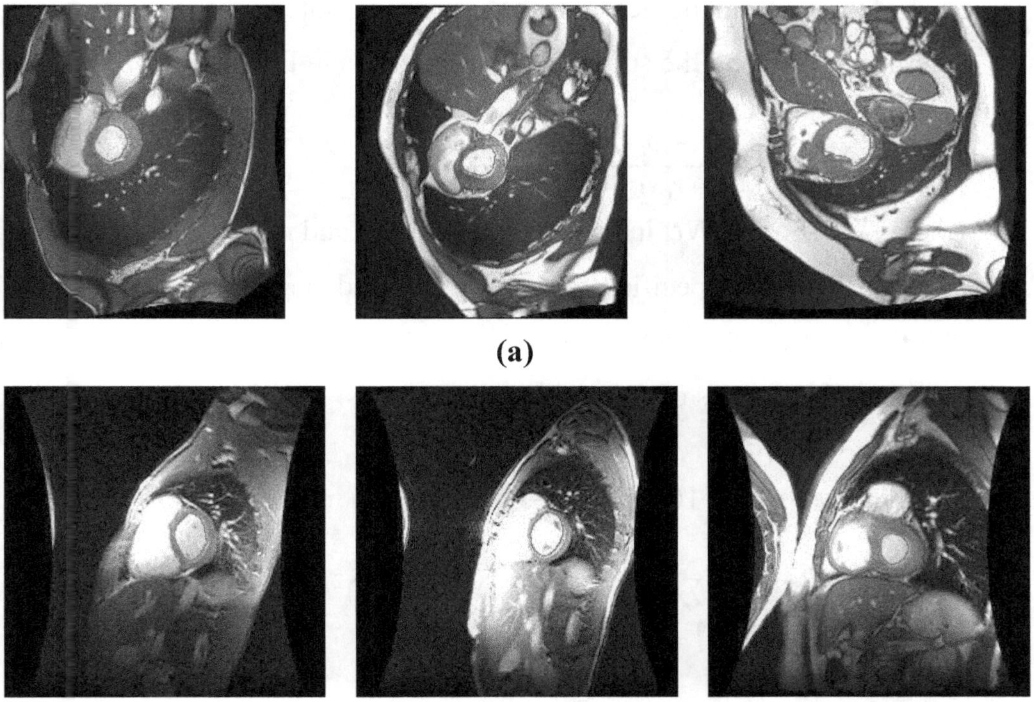

(a)

(b)

Figure 6.10 Sample segmentation results (a) Sunnybrook dataset and (b) York dataset

Figures 6.10 display the segmentation results for the proposed Tri-ConvNets models. The images are pre-processed and taken as input to the planned network. These inputs are processed in each convolutional layer. The convolutional layers are utilized for the LV segmentation, which is shown in above figure. The Figure 6.10 (a) demonstrates the segmentation result from the sunny brook dataset and Figure 6.10 (b) shows the segmentation result of LV from York dataset, respectively.

6.8 EVALUATION METRICS FOR TRI-CONVNETS

Analyzing various characteristics, such as accuracy, PSNR, MSE, DI and JI the results of the research demonstrate the dependability of the LV segmentation. Standard parameters such as True Positive (T_rPt^+), True Negative (T_rNt^-), False Positive (F_sPt^+), and False Negative (F_sNt^-) can be used to offer the assessment measures that have been stated. The performance has been determined utilising the subsequent statement as in Equation (6.7).

$$A = \frac{T_rPt^+ + T_rNt^-}{T_rPt^+ + T_rNt^- + F_sPt^+ + F_sNt^-} \times 100 \quad (6.7)$$

where, T_rPt^+ andT_rNt^+indicates true positives and negatives of the input images, $F_sPt^+$and F_sNt^- specifies false positives and negatives of the LV images. PSNR as in Equation (6.8), (6.9) and (6.10) is a data utilized in image analysis to rate the effectiveness of enhancement of images.

$$PSNR = 10 \log_{10} \left(\frac{r^2}{mse}\right) \quad (6.8)$$

$$MSE = \frac{1}{pq}\sum_{t=1}^{q}\sum_{f=1}^{p} (y(f,t) - z(f,t))^2 \quad (6.9)$$

$$MAE = \sum_{o=1}^{p} \frac{|g_o - \hat{g}|}{p} \quad (6.10)$$

The DI as in Equation (6.11) is a common metric to evaluate the accuracy of medical image segmentation algorithms, used for macular edema segmentation. It measures the overlap between the segmented region in the image, providing high quality of image with better segmentation accuracy. The statistical method used to evaluate the range and consistency in data sources is the JI as in Equation (6.12).

$$DI = \frac{2T_{pos}}{F_{pos} + 2T_{pos} + F_n} \quad (6.11)$$

$$JI = \frac{T_{pos}}{T_{pos}+F_{ng}+F_{pos}} \quad (6.12)$$

True positives and negatives of the sample images are denoted by T_{pos} and T_{ng}, while false positives and negatives are indicated by F_{pos} and F_{ng}.

6.9 COMPARISON ANALYSIS FOR PRE-PROCESSING FILTERS

In this second phase, several DL based filters are compared with traditional approaches with specific metrics include PSNR, MSE, and MAE, respectively. Table.6.7. depicts the comparison between denoising models.

Table 6.7 Analysis of the classic denoising filters performance comparisons

Filter	PSNR	MSE	MAE
Mean	31.25	0.28	0.89
Median	39.17	0.69	68.03
Histogram Equalization	45.22	0.71	64.83
Bilateral filter	46.32	0.83	67.42
Proposed ABIHE filter	52.21	0.98	70.31

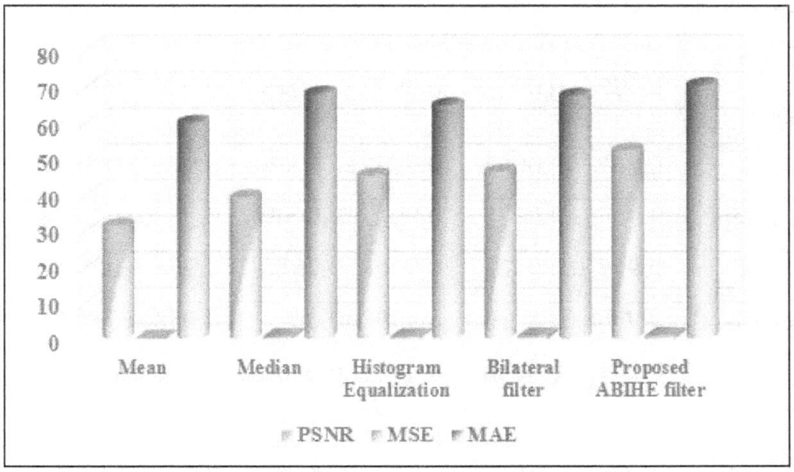

Figure 6.11 Comparison of classic denoising filters

Table 6.7 displays a comparison of the mean, median, Histogram Equalization, Bilateral filter, and ABIHE filter, as well as other traditional denoising filters, in terms of performance analysis. A minimal quantity of error rates was obtained by the proposed ABIHE filter. On the other hand, PSNR values are obtained higher than the other values. Comparing the ABIHE filter to different denoising strategies, the analysis found that it offers the least MSE. According to this conclusion, the ABIHE filters have the lowest MSE values, and when compared to other denoising filters, the suggested Tri-ConvNets exhibits low error rates for various noise ratios. Figure 6.11. visualizes the comparison of denoising filters.

6.10 COMPARISON ANALYSIS ON YORK DATASET

The suggested Tri-ConvNets segmentation approach outperforms other segmentation techniques, according to the performance metrics. The superior LV segmentation performance from the Tri-ConvNets approach is indicated by a higher PPV of 0.9880 and greater Jaccard index of 0.9497.

Table 6.8 Segmentation accuracy of York dataset with MR images

Algorithm	Sensitivity	Specificity	PPV	NPV
Tri-ConvNets	0.9792±0.0017	0.9936±0.0001	0.9880±0.0071	0.9992±0.0071
CNN+U-Net	0.9726±0.0025	0.9971±0.0019	0.9819±0.0106	0.9989±0.0106
CPR	0.9662±0.0062	0.9952±0.0030	0.9761±0.0142	0.9985±0.0142
MA-Shape	0.9643±0.0175	0.9923±0.0009	0.9700±0.0178	0.9981±0.0178
DLDP	0.9577±0.0123	0.9908±0.0086	0.9640±0.0211	0.9977±0.0211
Omega-Net	0.9508±0.0216	0.9900±0.0032	0.9584±0.0251	0.9973±0.0251
FCN	0.9432±0.0230	0.9941±0.0062	0.9519±0.0284	0.9969±0.0284

Table 6.9 York dataset segmentation validation with dice and Jaccard indices

Algorithm	Dice Index	Jaccard Index
Tri-ConvNets	0.9403±0.0071	0.9497±0.0071
CNN+U-Net	0.9365±0.0106	0.9496±0.0106
CPR	0.9333±0.0142	0.9480±0.0142
MA-Shape	0.9332±0.0178	0.9479±0.0178
DLDP	0.9315±0.0211	0.9476±0.0211
Omega-Net	0.9322±0.0251	0.9475±0.0251
FCN	0.9323±0.0284	0.9466±0.0284

Figure 6.12 illustrates the higher similarity between the automated and the manual segmentation on York dataset MR images. Figure 6.10 depicts sample segmentation results of short axis MR images from all four datasets.

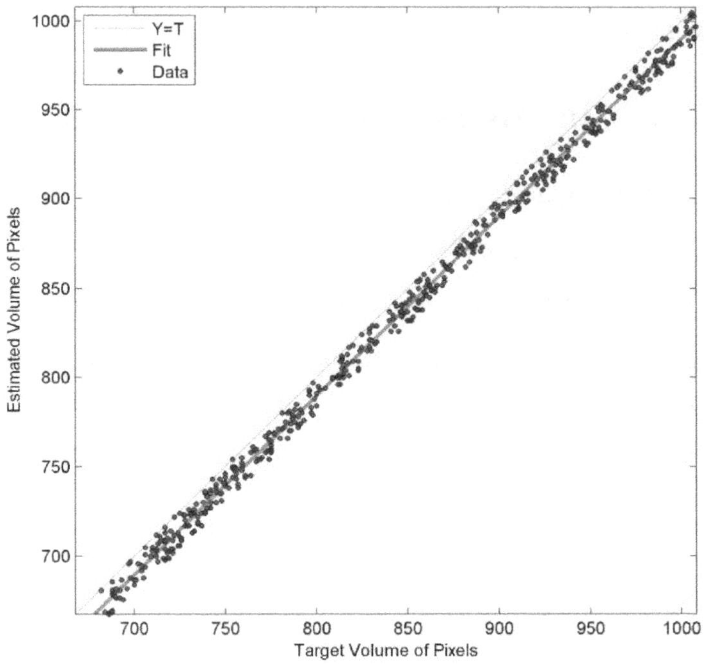

Figure 6.12 Linear regression plot depicting the automatic-segmentation volumes against manually obtained volumes with MR images from york dataset

Table 6.10 Performance comparison of existing techniques with proposed model

Author	Methods	Dice Index	Jaccard Index
Zhang *et al.* (2022)	FC-Dense Net	0.9466±0.0120	0.9415±0.0110
Gayathri *et al.* (2023)	ES-FCN	0.9427±0.0120	0.9449±0.0120
Proposed model	Tri-ConvNets	0.9403±0.0071	0.9457±0.0071

Table 6.10 quantifies the segmentation performance measures estimated of the LV images from the York dataset. The average and variance values of the performance measures are depicted. The experimental result indicates that the proposed Tri-ConvNets segmentation approach performs better than the other models with higher accuracy. The greater *PPV* of 0.9582 and higher Jaccard index of 0.9477 indicates the effective and sensitive LV segmentation from the Tri-ConvNets method.

6.11 COMPARISON ANALYSIS ON SUNNYBROOK DATASET

The suggested Tri-ConvNets approach has the ability to segment the LV regions more effectively than the other techniques, according to its effectiveness criteria. The increased Jaccard index of 0.9491 and enhanced PPV of 0.9861 demonstrate the Tri-ConvNets technique's noteworthy LV segmentation ability.

Table 6.11 Comparison of segmentation accuracy with MR images from sunnybrook dataset

Algorithm	Sensitivity	Specificity	PPV	NPV
Tri-ConvNets	0.9832±0.0068	0.9895±0.0007	0.9861±0.0080	0.9992±0.0080
CNN+U-Net	0.9834±0.0023	0.9931±0.0083	0.9844±0.0090	0.9991±0.0090
CPR	0.9788±0.0085	0.9928±0.0008	0.9826±0.0101	0.9990±0.0101
MA-Shape	0.9715±0.0052	0.9904±0.0046	0.9810±0.0111	0.9989±0.0111
DLDP	0.9740±0.0098	0.9954±0.0073	0.9793±0.0120	0.9988±0.0120
Omega-Net	0.9710±0.0078	0.9960±0.0027	0.9775±0.0130	0.9988±0.0130
FCN	0.9722±0.0077	0.9915±0.0056	0.9757±0.0140	0.9986±0.0140

Table 6.12 Sunnybrook dataset segmentation estimation with dice and jaccard indices

Algorithm	Dice Index	Jaccard Index
Tri-ConvNets	0.9389±0.0080	0.9491±0.0080
CNN+U-Net	0.9376±0.0090	0.9489±0.0090
CPR	0.9371±0.0101	0.9490±0.0101
MA-Shape	0.9358±0.0111	0.9489±0.0111
DLDP	0.9349±0.0120	0.9487±0.0120
Omega-Net	0.9344±0.0130	0.9489±0.0130
FCN	0.9330±0.0140	0.9486±0.0140

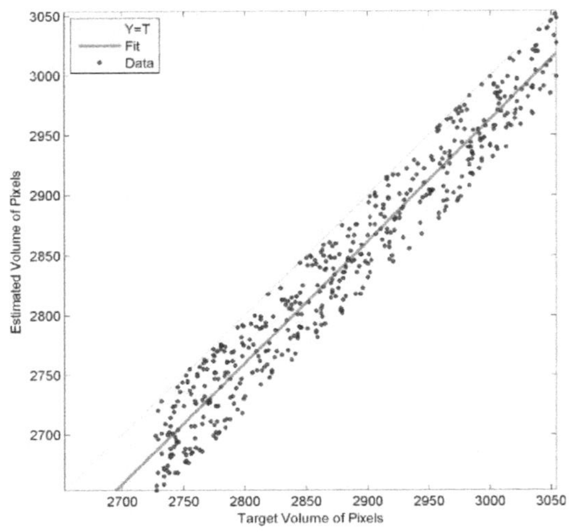

Figure 6.13 Linear regression plot depicting the automatic-segmentation volumes against manually obtained volumes with MR images from sunnybrook dataset

Table 6.11 & 6.12 listed the segmentation accuracy from Sunnybrook dataset. Figure 6.13 illustrates the higher similarity between the automated and the manual segmentation on Sunnybrook dataset MR images.

Table 6.13 Performance comparison of existing techniques with proposed model

Author	Methods	Dice Index	Jaccard Index
Seo *et al.* (2019)	2D U-Net	0.9451±0.0120	0.9435±0.0110
Habijan *et al.* (2019)	CNN	0.9434±0.0120	0.9469±0.0120
Bekouche *et al.* (2023)	U-Net	0.9434±0.0120	0.9431±0.0120
Proposed model	Tri-ConvNets	0.9389±0.0080	0.9491±0.0080

Table 6.13 quantifies the segmentation performance measures estimated of the LV images from the York dataset. The average and variance values of the performance measures are depicted. The experimental result indicates that the proposed Tri-ConvNets segmentation approach performs better

than the other models with higher accuracy. The greater PPV of 0.9582 and higher Jaccard index of 0.9477 indicates the effective and sensitive LV segmentation from the Tri-ConvNets method.

6.12 SUMMARY

In this chapter, the four-dataset used for both the proposed algorithms is elaborated, and the dataset details are mentioned in the table. The Evaluation metrics are also explained. The performance of the proposed algorithm is compared with other algorithms and explained in the table in this chapter.

A CERTAIN INVESTIGATION ON HEART DISEASES DETECTION SYSTEM USING DEEP LEARNING TECHNIQUES

A THESIS

Submitted by

GAYATHRI K

in partial fulfillment of the requirements for the degree of

DOCTOR OF PHILOSOPHY

FACULTY OF INFORMATION AND COMMUNICATION ENGINEERING

ANNA UNIVERSITY

CHENNAI 600 025

OCTOBER 2024

CHAPTER 7

CONCLUSION AND FUTURE WORK

Segmentation of the Left Ventricle (LV) region from the short-axis MR image is focused here in this research work. An efficient Fully Convolutional Neural Network (FCN) framework based on edge and shape features is proposed for more accurate segmentation. The segmentation framework uses binary images as input instead of grayscale or color images as the other baseline and existing segmentation method employs. Processing the binary images and shape-based FCN weight manipulation accelerates the segmentation procedure, and also achieves more exact segmentation results which are closely related to the manual segmentation.

In the First Phase the proposed Edge and Shape-based FCN (ES-FCN) segmentation method is evaluated with six performance metrics and analyzed with six existing segmentation algorithms. The ACDC and LVSC datasets were used to evaluate the proposed segmentation methodology and produced Jaccard indices of 0.9484±0.0188 and 0.9476±0.0237, respectively. The experimental results indicate superior and significant segmentation performance with higher Positive and Negative Predictivevalues. Although the proposed ES-FCN model performs admirably in LV segmentation, it has a drawback that makes this method less effective.

A combination of different deep neural networks for semantic segmentation of the left ventricle based on Tri-ConvNetsis also proposed in Second Phase to obtain accurate segmentation. CMRI images are initially pre-processed to remove noise artefacts and enhance image quality, then ROI-based extraction is done in three stages to accurately identify the LV. The extracted features are given as input to three different deep learning structures for segmenting the LV in an efficient way. The contour edges are processed in the standard ConvNet, the contour points are processed using Fully ConvNet and finally the noise free images are converted into patches to perform pixel-wise operations in ConvNets. These three ConvNets perform different operations to give different segmentation results and these results concatenate to calculate circular similarity for attaining better LV segmentation output. The proposed Tri-ConvNets model achieves the Jaccard indices of 0.9491±0.0188 for the sunny brook dataset and 0.9497±0.0237 for the York dataset, and the dice index of 0.9419±0.0178 for the ACDC dataset and 0.9414±0.0247 for LVSC dataset, respectively.

In LV segmentation, the major class imbalance problem raises false negative rates, which will be addressed in future work. Further, medical imaging will be used in the future to assess the proposed model for reducing the need for manual segmentation by clinical experts during cardiac diagnosis. In the future, more medical imaging will be used to assess the proposed model. In future researchers enhance model interpretability and generalization by integrating attention mechanisms and self-supervised learning. Exploring multi-modal fusion to improve the robustness and accuracy of left ventricle segmentation for clinical applications.

REFERENCES

1. Abdeltawab, H, Khalifa, F & Taher, F 2020, 'A deep learning-based approach for automatic segmentation and quantification of the left ventricle from cardiac cine MR images', Computerized Medical Imaging and Graphics, vol. 81, Article No: 101717.

2. Amer, A, Ye, X & Janan, F 2021, 'ResDUnet: A deep learning-based left ventricle segmentation method for echocardiography', IEEE Access, vol. 9, pp. 159755-159763.

3. Amer, A, Ye, X, Zolgharni, M & Janan, F 2020, 'ResDUnet: Residual Dilated UNet for left ventricle segmentation from echocardiographic images', in 2020 42nd Annual International Conference of the IEEE Engineering in Medicine & Biology Society (EMBC), IEEE, pp. 2019-2022.

4. Avendi, MR, Kheradvar, A & Jafarkhani, H 2016, 'A combined deep-learning and deformable-model approach to fully automatic segmentation of the left ventricle in cardiac MRI', Medical Image Analysis, vol. 30, pp. 108-119.

5. Bai, W, Oktay, O, Sinclair, M, Suzuki, H, Rajchl, M, Tarroni, G & Rueckert, D 2017, 'Semi-supervised learning for network-based cardiac MR image segmentation', in International Conference on Medical Image Computing and Computer-Assisted Intervention, Springer, Cham, pp. 253-260.

6. Bai, W, Sinclair, M, Tarroni, G, Oktay, O, Rajchl, M, Vaillant, G & Rueckert, D 2018, 'Automated cardiovascular magnetic resonance image analysis with fully convolutional networks', Journal of Cardiovascular Magnetic Resonance, vol. 20, no. 65, pp. 1-12.

7. Bekkouche, A, Merzoug, M, Hadjila, F & Daoud, O 2023, 'Segmentation of the left ventricle using improved Unet neural networks' Preprints 2023, Available from: <https://doi.org/10.20944/preprints202308.1719.v1>, pp. 1-13.

8. Bernard, O, Lalande, A, Zotti, C, Cervenansky, F, Yang, X, Heng, PA & Jodoin, PM 2018, 'Deep learning techniques for automatic MRI cardiac multi-structures segmentation and diagnosis: Is the problem solved?', IEEE Transactions on Medical Imaging, vol. 37, no. 11, pp. 2514-2525.

9. Budai, A, Suhai, FI, Csorba, K, Toth, A, Szabo, L, Vago, H & Merkely, B 2020, 'Fully automatic segmentation of right and left ventricle on short-axis cardiac MRI images', Computerized Medical Imaging and Graphics, vol. 85, Article No: 101786 .

10. Chen, C, Qin, C, Qiu, H, Tarroni, G, Duan, J, Bai, W & Rueckert, D 2020, 'Deep learning for cardiac image segmentation: A review', Frontiers in Cardiovascular Medicine, vol. 7, pp. 1-35.

11. Choromanska, A, Henaff, M, Mathieu, M, Arous, GB & LeCun, Y 2015, 'The loss surfaces of multilayer networks', in Artificial Intelligence and Statistics, PMLR, vol. 38, pp. 192-204.

12. Cong, J, Zheng, Y, Xue, W, Cao, B & Li, S 2019, 'Ma-shape: Modality adaptation shape regression for left ventricle segmentation on mixed MR and CT images', IEEE Access, vol. 7, pp. 16584-16593.

13. Counseller, Q & Aboelkassem, Y 2023, 'Recent technologies in cardiac imaging', Frontiers in Medical Technology, vol. 4, pp. 1-17.

14. Csecs, I, Czimbalmos, C, Suhai, FI, Mikle, R, Mirzahosseini, A, Dohy, Z & Vágó, H 2018, 'Left and right ventricular parameters corrected with threshold-based quantification method in a normal cohort analyzed by three independent observers with various training-degree', The International Journal of Cardiovascular Imaging, vol. 34, no. 7, pp. 1127-1133.

15. Cui, H, Yuwen, C, Jiang, L, Xia, Y & Zhang, Y 2021, 'Multiscale attention guided U-net architecture for cardiac segmentation in short- axis MRI images', Computer Methods and Programs in Biomedicine, vol. 206, Article No: 106142.

16. Curiale, AH, Colavecchia, FD & Mato, G 2019, 'Automatic quantification of the LV function and mass: A deep learning approach for cardiovascular MRI', Computer Methods and Programs in Biomedicine, vol. 169, pp. 37-50.

17. Dreijer, JF, Herbst, BM & Du Preez, JA 2013, 'Left ventricular segmentation from MRI datasets with edge modelling conditional random fields', BMC Medical Imaging, vol. 13, Article No: 24, pp. 1-23.

18. Duan, J, Bello, G, Schlemper, J, Bai, W, Dawes, TJ, Biffi, C & Rueckert, D 2019, 'Automatic 3D bi-ventricular segmentation of cardiac images by a shape-refined multi-task deep learning approach', IEEE Transactions on Medical Imaging, vol. 38, no. 9, pp. 2151-2164.

19. Fenil, E, Manogaran, G, Vivekananda, GN, Thanjaivadivel, T, Jeeva, S & Ahilan, AJCN 2019, 'Real time violence detection framework for football stadium comprising of big data analysis and deep learning through bidirectional LSTM', Computer Networks, vol. 151, pp. 191-200.

20. Folkesson, J, Samset, E, Kwong, RY & Westin, CF 2008, 'Unifying statistical classification and geodesic active regions for segmentation of cardiac MRI', IEEE Transactions on Information Technology in Biomedicine, vol. 12, no. 3, pp. 328-334.

21. Gayathri, K, Ajitha Gladis, KP & Angel Mary, A 2023, 'Real time masked face recognition using deep learning based yolov4 network', International Journal of Data Science and Artificial Intelligence, vol. 01, no. 01, pp. 26-32.

22. Gayathri, K, Uma Maheswari, N, Venkatesh, R & Appathurai, A 2023 'Automatic left ventricle segmentation via edge-shape feature-based fully convolutional neural network', International Journal of Imaging Systems and Technology, vol. 34, no. 1, pp. 1-18.

23. Glorot, X & Bengio, Y 2010, 'Understanding the difficulty of training deep feedforward neural networks', in Proceedings of the Thirteenth International Conference on Artificial Intelligence and Statistics, JMLR Workshop and Conference Proceedings, vol. 9, pp. 249-256.

24. Graves, CV, Moreno, RA, Rebelo, MS, Nomura, CH & Gutierrez, MA 2020, 'Improving the generalization of deep learning methods to segment the left ventricle in short axis MR images', in 2020 42nd Annual International Conference of the IEEE Engineering in Medicine & Biology Society (EMBC), IEEE, pp. 1203-1206.

25. Grosgeorge, D, Petitjean, C, Dacher, JN & Ruan, S 2013, 'Graph cut segmentation with a statistical shape model in cardiac MRI', Computer Vision and Image Understanding, vol. 117, no. 9, pp. 1027-1035.

26. Grothues, F, Moon, JC, Bellenger, NG, Smith, GS, Klein, HU & Pennell, DJ 2004, 'Interstudy reproducibility of right ventricular volumes, function, and mass with cardiovascular magnetic resonance', American Heart Journal, vol. 147, no. 2, pp. 218-223.

27. Gu, J, Wang, Z, Kuen, J, Ma, L, Shahroudy, A, Shuai, B & Chen, T 2018, 'Recent advances in convolutional neural networks', Pattern Recognition, vol. 77, pp. 354-377.

28. Habijan, M, Leventić, H, Galić, I & Babin, D 2019, 'Estimation of the left ventricle volume using semantic segmentation', in 2019 International Symposium ELMAR, IEEE, pp. 39-44.

29. Hadhoud, MM, Eladawy, MI, Farag, A, Montevecchi, FM & Morbiducci, U 2012, 'Left ventricle segmentation in cardiac MRI images', American Journal of Biomedical Engineering, vol. 2, no. 3, pp. 131-135.

30. Halliday, BP, Senior, R & Pennell, DJ 2021, 'Assessing left ventricular systolic function: From ejection fraction to strain analysis', European Heart Journal, vol. 42, no. 7, pp. 789-797.

31. Hassan, SA, Palaskas, NL, Agha, AM, Iliescu, C, Lopez-Mattei, J, Chen, C, Zheng, H & Yusuf, SW 2019, 'Carcinoid heart disease: A comprehensive review', Current Cardiology Reports, vol. 21, pp. 1-7.

32. He, K, Zhang, X, Ren, S & Sun, J 2015, 'Delving deep into rectifiers: Surpassing human-level performance on imagenet classification', in Proceedings of the IEEE International Conference on Computer Vision, pp. 1026-1034.

33. Hsu, WY 2019, 'Automatic left ventricle recognition, segmentation and tracking in cardiac ultrasound image sequences', IEEE Access, vol. 7, pp. 140524-140533.

34. Hu, H, Gao, Z, Liu, L, Liu, H, Gao, J, Xu, S & Huang, L 2014, 'Automatic segmentation of the left ventricle in cardiac MRI using local binary fitting model and dynamic programming techniques', PloS One, vol. 9, no. 12, pp. 1-17.

35. Hu, H, Pan, N, Wang, J, Yin, T & Ye, R 2019, 'Automatic segmentation of left ventricle from cardiac MRI via deep learning and region constrained dynamic programming', Neurocomputing, vol. 347, pp. 139-148.

36. Hubel, DH & Wiesel, TN 1959, 'Receptive fields of single neurones in the cat's striate cortex', The Journal of Physiology, vol. 148, no. 3, pp. 574-591.

37. Irshad, M, Yasmin, M, Sharif, MI, Rashid, M, Sharif, MI & Kadry, S 2023, 'A novel light U-Net model for left ventricle segmentation using MRI', Mathematics, vol. 11, no. 14, pp. 1-21.

38. Jang, Y, Hong, Y, Ha, S, Kim, S & Chang, HJ 2017, 'Automatic segmentation of LV and RV in cardiac MRI', in International Workshop on Statistical Atlases and Computational Models of the Heart, Springer, Cham, vol. 10663, pp. 161-169.

39. Joyce, T, Buoso, S, Stoeck, CT & Kozerke, S 2022, 'Rapid inference of personalised left-ventricular meshes by deformation-based differentiable mesh voxelization', Medical Image Analysis, vol. 79, Article No: 102445.

40. Kaus, MR, Von Berg, J, Weese, J, Niessen, W & Pekar, V 2004, 'Automated segmentation of the left ventricle in cardiac MRI', Medical Image Analysis, vol. 8, no. 3, pp. 245-254.

41. Khened, M, Kollerathu, VA & Krishnamurthi, G 2019, 'Fully convolutional multi-scale residual DenseNets for cardiac segmentation and automated cardiac diagnosis using ensemble of classifiers', Medical Image Analysis, vol. 51, pp. 21-45.

42. Leclerc, S, Smistad, E, Østvik, A, Cervenansky, F, Espinosa, F, Espeland, T, Berg, EAR, Belhamissi, M, Israilov, S, Grenier, T & Lartizien, C 2020, 'LU-Net: A multistage attention network to improve the robustness of segmentation of left ventricular structures in 2-D echocardiography', IEEE Transactions on Ultrasonics, Ferroelectrics, and Frequency Control, vol. 67, no. 12, pp. 2519-2530.

43. LeCun, YA, Bottou, L, Orr, GB & Müller, KR 2012, 'Efficient backprop', in Neural Networks: Tricks of the Trade, Springer, Berlin, Heidelberg, vol. 7700, pp. 9-48.

44. Li, H, Wang, Y, Qu, M, Cao, P, Feng, C & Yang, J 2023, 'EchoEFNet: Multi-task deep learning network for automatic calculation of left ventricular ejection fraction in 2D echocardiography', Computers in Biology and Medicine, vol. 156, Article No: 106705.

45. Lin, A, Wu, J & Yang, X 2020, 'A data augmentation approach to train fully convolutional networks for left ventricle segmentation', Magnetic Resonance Imaging, vol. 66, pp.152-164.

46. Lin, M, Chen, Q & Yan, S 2013, 'Network in network', arXiv Preprint Arxiv, Available from: <https://doi.org/10.48550/arXiv.1312.4400>, pp. 1-10.

47. Long, J, Shelhamer, E & Darrell, T 2015, 'Fully convolutional networks for semantic segmentation', in Proceedings of the IEEE Conference on Computer Vision and Pattern Recognition, pp. 3431-3440.

48. Mahapatra, D 2013, 'Cardiac image segmentation from cine cardiac MRI using graph cuts and shape priors', Journal of Digital Imaging, vol. 26, no. 4, pp. 721-730.

49. Margeta, J, Geremia, E, Criminisi, A & Ayache, N 2011, 'Layered spatio-temporal forests for left ventricle segmentation from 4D cardiac MRI data', in International Workshop on Statistical Atlases and Computational Models of the Heart, Springer, Berlin, Heidelberg, vol. 7085, pp. 109-119.

50. Miller, CA, Jordan, P, Borg, A, Argyle, R, Clark, D, Pearce, K & Schmitt, M 2013, 'Quantification of left ventricular indices from SSFP cine imaging: Impact of real-world variability in analysis methodology and utility of geometric modeling', Journal of Magnetic Resonance Imaging, vol. 37, no. 5, pp. 1213-1222.

51. Mortazi, A & Bagci, U 2018, 'Automatically designing CNN architectures for medical image segmentation', in International Workshop on Machine Learning in Medical Imaging, Springer, Cham, vol. 11046, pp. 98-106.

52. Nair, V & Hinton, GE 2010, 'Rectified linear units improve restricted boltzmann machines', in Proceedings of the 27th International Conference on International Conference on Machine Learning, pp. 807-814.

53. Nambakhsh, CM, Yuan, J, Punithakumar, K, Goela, A, Rajchl, M, Peters, TM & Ayed, IB 2013, 'Left ventricle segmentation in MRI via convex relaxed distribution matching', Medical Image Analysis, vol. 17, no. 8, pp. 1010-1024.

54. Ngo, TA & Carneiro, G 2013, 'Left ventricle segmentation from cardiac MRI combining level set methods with deep belief networks', in 2013 IEEE International Conference on Image Processing, IEEE, pp. 695-699.

55. Painchaud, N, Skandarani, Y, Judge, T, Bernard, O, Lalande, A & Jodoin, PM 2019, 'Cardiac MRI segmentation with strong anatomical guarantees', in International Conference on Medical Image Computing and Computer-Assisted Intervention, Springer, Cham, vol. 11765, pp. 632-640.

56. Pastore, MC, Mandoli, GE, Aboumarie, HS, Santoro, C, Bandera, F, D'Andrea, A, Benfari, G, Esposito, R, Evola, V, Sorrentino, R & Cameli, P 2020, 'Basic and advanced echocardiography in advanced heart failure: An overview', Heart Failure Reviews, vol. 25, pp. 937-948.

57. Patravali, J, Jain, S & Chilamkurthy, S 2017, '2D-3D fully convolutional neural networks for cardiac MR segmentation', in International Workshop on Statistical Atlases and Computational Models of the Heart, Springer, Cham, vol. 10663, pp. 130-139.

58. Pednekar, A, Kurkure, U, Muthupillai, R, Flamm, S & Kakadiaris, IA 2006, 'Automated left ventricular segmentation in cardiac MRI', IEEE Transactions on Biomedical Engineering, vol. 53, no. 7, pp. 1425-1428.

59. Petitjean, C & Dacher, JN 2011, 'A review of segmentation methods in short axis cardiac MR images', Medical Image Analysis, vol. 15, no. 2, pp. 169-184.

60. Pham, VT & Tran, TT 2016, 'Active contour model and nonlinear shape priors with application to left ventricle segmentation in cardiac MR images', Optik, vol. 127, no. 3, pp. 991-1002.

61. Pluempitiwiriyawej, C, Moura, JM, Wu, YJL & Ho, C 2005, 'STACS: New active contour scheme for cardiac MR image segmentation', IEEE Transactions on Medical Imaging, vol. 24, no. 5, pp. 593-603.

62. Poudel, RP, Lamata, P & Montana, G 2016, 'Recurrent fully convolutional neural networks for multi-Slice MRI cardiac segmentation', in Reconstruction, Segmentation, and Analysis of Medical Images, Springer, Cham, vol. 10129, pp. 83- 94.

63. Queirós, S, Barbosa, D, Heyde, B, Morais, P, Vilaça, JL, Friboulet, D & D'hooge, J 2014, 'Fast automatic myocardial segmentation in 4D cine CMR datasets', Medical Image Analysis, vol. 18, no. 7, pp. 1115-1131.

64. Radau, P, Lu, Y, Connelly, K, Paul, G, Dick, A & Wright, G 2009, 'Evaluation framework for algorithms segmenting short axis cardiac MRI', The MIDAS Journal-Cardiac MR Left Ventricle Segmentation Challenge, vol. 49, pp. 1-7.

65. Rashid, S, Qureshi, AG, Noor, TA, Yaseen, K, Sheikh, MAA, Malik, M & Malik, J 2023, 'Anxiety and depression in heart failure: An updated review', Current Problems in Cardiology, vol. 48, no. 11, P. 101987.

66. Ribeiro, MA & Nunes, FL 2022, 'Left ventricle segmentation in cardiac MR: A systematic mapping of the past decade', ACM Computing Surveys (CSUR), vol. 54, no. 11, pp. 1-38.

67. Ripley, DP, Musa, TA, Dobson, LE, Plein, S & Greenwood, JP 2016, 'Cardiovascular magnetic resonance imaging: What the general cardiologist should know', Heart, vol. 102, no. 19, pp. 1589-1603.

68. Robinson, R, Valindria, VV, Bai, W, Suzuki, H, Matthews, PM, Page, C & Glocker, B 2017, 'Automatic quality control of cardiac MRI segmentation in large-scale population imaging', in International Conference on Medical Image Computing and Computer-Assisted Intervention, Springer, Cham, vol. 10433, pp. 720-727.

69. Rostami, A, Amirani, MC & Yousef-Banaem, H 2020, 'Segmentation of the left ventricle in cardiac MRI based on convolutional neural network and level set function', Health Technology, vol. 10, no. 5, pp. 1155-1162.

70. Russakovsky, O, Deng, J, Su, H, Krause, J, Satheesh, S, Ma, S & Fei-Fei, L 2015, 'Imagenet large scale visual recognition challenge', International Journal of Computer Vision, vol. 115, no. 3, pp. 211-252.

71. Sander, J, de Vos, BD, Wolterink, JM & Išgum, I 2019, 'Towards increased trustworthiness of deep learning segmentation methods on cardiac MRI', in Medical Imaging 2019: Image Processing, International Society for Optics and Photonics, vol. 10949, pp. 1-8.

72. Santiago, C, Nascimento, JC & Marques, JS 2017, 'A new ASM framework for left ventricle segmentation exploring slice variability in cardiac MRI volumes', Neural Computing and Applications, vol. 28, no. 9, pp. 2489-2500.

73. Santiago, C, Nascimento, JC & Marques, JS 2018, 'Fast segmentation of the left ventricle in cardiac MRI using dynamic programming', Computer Methods and Programs in Biomedicine, vol. 154, pp. 9-23.

74. Seo, B, Mariano, D, Beckfield, J, Madenur, V, Hu, Y, Reina, T, Bobar, M, Nguyen, MH & Altintas, I 2019, 'Cardiac MRI image segmentation for left ventricle and right ventricle using deep learning', arXiv Preprint, Available from: <https://doi.org/10.48550/arXiv.1909.08028>, pp. 1-27.

75. Shi, X & Li, C 2021, 'Convexity preserving level set for left ventricle segmentation,' Magnetic Resonance Imaging, vol. 78, pp.109-118.

76. Sivasankari, B, Shunmugathammal, M, Appathurai, A & Kavitha, M 2022, 'High-throughput and power-efficient convolutional neural network using one-pass processing elements', Journal of Circuits, Systems and Computers, vol. 31, no. 13, Article No: 2250226.

77. Steinhaus, DA, Lubitz, SA, Noseworthy, PA & Kramer, DB 2019, 'Exercise interventions in patients with implantable cardioverter-defibrillators and cardiac resynchronization therapy: A systematic review and meta-analysis', Journal of Cardiopulmonary Rehabilitation and Prevention, vol. 39, no. 5, pp. 308-317.

78. Suinesiaputra, A, Cowan, BR, Al-Agamy, AO, Elattar, MA, Ayache, N, Fahmy, AS & Young, AA 2014, 'A collaborative resource to build consensus for automated left ventricular segmentation of cardiac MR images', Medical Image Analysis, vol. 18, no. 1, pp. 50-62.

79. Sun, H, Frangi, AF, Wang, H, Sukno, FM, Tobon-Gomez, C & Yushkevich, PA 2010, 'Automatic cardiac MRI segmentation using a biventricular deformable medial model', in International Conference on Medical Image Computing and Computer-Assisted Intervention, Springer, Berlin, Heidelberg, pp. 468-475.

80. Sundarasekar, R & Appathurai, A 2021, 'Efficient brain tumor detection and classification using magnetic resonance imaging', Biomedical Physics Engineering Express, vol. 7, no. 5.

81. Tan, LK, McLaughlin, RA, Lim, E, Abdul Aziz, YF & Liew, YM 2018, 'Fully automated segmentation of the left ventricle in cine cardiac MRI using neural network regression', Journal of Magnetic Resonance Imaging, vol. 48, no. 1, pp. 140-152.

82. Tavakoli, V & Amini, AA 2013, 'A survey of shaped-based registration and segmentation techniques for cardiac images', Computer Vision and Image Understanding, vol. 117, no. 9, pp. 966-989.

83. Teixeira, GM, Pommeranzembaum, IR, de Oliveira, BL, Lobosco, M & Dos Santos, RW 2008, 'Automatic segmentation of cardiac MRI using snakes and genetic algorithms', in International Conference on Computational Science, Springer, Berlin, Heidelberg, pp. 168-177.

84. Tran, PV 2016, 'A fully convolutional neural network for cardiac segmentation in short-axis MRI', arXiv preprint, Available from: <https://doi.org/10.48550/arXiv.1604.00494>, pp. 1-21.

85. Vesal, S, Ravikumar, N & Maier, A 2019, 'Automated multi-sequence cardiac MRI segmentation using supervised domain adaptation', in International Workshop on Statistical Atlases and Computational Models of the Heart, Springer, Cham, pp. 300-308.

86. Vigneault, DM, Xie, W, Ho, CY, Bluemke, DA & Noble, JA 2018, 'Ω-net (Omega-net): Fully automatic, multi-view cardiac MR detection, orientation, and segmentation with deep neural networks', Medical Image Analysis, vol. 48, pp. 95-106.

87. Wang, Z, Peng, Y, Li, D, Guo, Y & Zhang, B 2022, 'MMNet: A multi-scale deep learning network for the left ventricular segmentation of cardiac MRI images', Apple Intelligent, vol. 52, no. 5, pp. 5225-5240.

88. Wang, Z, Xie, L & Qi, J 2022, 'Dynamic pixel-wise weighting-based fully convolutional neural networks for left ventricle segmentation in short-axis MRI', Magnetic Reson Imaging, vol. 66, pp. 131-140.

89. Wijnhoven, RG & de With, PHN 2010, 'Fast training of object detection using stochastic gradient descent', in 2010 20th International Conference on Pattern Recognition, IEEE, pp. 424-427.

90. Woo, J, Slomka, PJ, Kuo, CCJ & Hong, BW 2013, 'Multiphase segmentation using an implicit dual shape prior: Application to detection of left ventricle in cardiac MRI', Computer Vision and Image Understanding, vol. 117, no. 9, pp. 1084-1094.

91. Wu, B, Fang, Y & Lai, X 2020, 'Left ventricle automatic segmentation in cardiac MRI using a combined CNN and U-net approach', Computerized Medical Imaging and Graphics, vol. 82, Article No: 101719.

92. Wu, H, Lu, X, Lei, B & Wen, Z 2021, 'Automated left ventricular segmen- tation from cardiac magnetic resonance images via adversarial learning with multi-stage pose estimation network and co-dis-criminator', Medical Image Analysis, vol. 68, Article No: 101891.

93. Xie, L, Song, Y & Chen, Q 2020, 'Automatic left ventricle segmentation in short-axis MRI using deep convolutional neural networks and central-line guided level set approach', Compute Biology Medical, vol. 22, Article No: 103877.

94. Yang, M, Xiao, X, Liu, Z, Sun, L, Guo, W, Cui, L, Sun, D, Zhang, P & Yang, G 2020, 'Deep retinanet for dynamic left ventricle detection in multiview echocardiography classification', Scientific Programming, vol. 2020, Article No: 7025403, pp. 1-6.

95. Yu, D, Wang, H, Chen, P & Wei, Z 2014, 'Mixed pooling for convolutional neural networks', in International Conference on Rough Sets and Knowledge Technology, Springer, Cham, vol. 8818, pp. 364-375.

96. Zhang, J, Zhang, Y, Zhang, H, Zhang, Q, Su, W, Guo, S & Wang, Y 2022, 'Segmentation of biventricle in cardiac cine MRI via nested capsule dense network', Peer Journal Computer Science, vol. 8, P. e1146.

97. Zhu, Z, Khan, MA, Wang, SH & Zhang, YD 2023, 'RBEBT: A ResNet-based BA-ELM for brain tumor classification', Compute Materials, vol. 74, no. 1, pp. 101-111.

98. Zhu, Z, Ren, Z, Lu, S, Wang, S & Zhang, Y 2023, 'DLBCNet: A deep learning network for classifying blood cells', Big Data Cognitive Computing, vol. 7, no. 2, pp. 1-20.

99. Zhu, Z, Wang, SH & Zhang, YD 2022, 'A survey of convolutional neural network in breast cancer', Compute Model Engineering Science, vol. 136, no. 3, pp. 2127-2172.

100. Zhu, Z, Wang, SH & Zhang, YD 2023, 'ReRNet: a deep learning network for classifying blood cells', Technology in Cancer Research & Treatment, vol. 22, pp.1-14.

101. Zhuang, X, Hawkes, DJ, Crum, WR, Boubertakh, R, Uribe, S, Atkinson, D & Hill, DL 2008, 'Robust registration between cardiac MRI images and atlas for segmentation propagation', in Medical Imaging 2008: Image Processing, International Society for Optics and Photonics, vol. 6914, Article No: 691408.

102. Zinkevich, MA, Weimer, M, Smola, A & Li, L 2010, 'Parallelized stochastic gradient descent', in Proceedings of the 23rd International Conference on Neural Information Processing Systems, vol. 2, pp. 2595-2603.

LIST OF PUBLICATIONS

International Journal

1. **Gayathri, K**, Uma Maheswari, N, Venkatesh, R & Appathurai, A 2023, 'Automatic left ventricle segmentation via edge-shape feature-based fully convolutional neural network', International Journal of Imaging Systems and Technology, vol. 34, no. 1, P. e22497, **Impact Factor: 4.05.**

2. **Gayathri, K,** Uma Maheswari, N, Venkatesh, R & Ganesh Prabu, B 2024, 'Fine grained automatic left ventricle segmentation via ROI based Tri-Convolutional neural networks', Technological Health Care, pp. 1-23, DOI: 10.3233/THC-240062.

www.ingramcontent.com/pod-product-compliance
Lightning Source LLC
LaVergne TN
LVHW082349200926
848481LV00012B/3505
9798231443536